# NATO

## STRUCTURE AND OPERATION

Juan Antonio Falcón Blasco

# CONTENTS

# PREFACE

The security environment in and around the world has become more complex and threats have multiplied. NATO wants to guarantee security for its citizens in an unstable world. This approach is also a response to the demand by the majority of the NATO citizens, who want more security, stability and a coordinated NATO response to current threats.

As a summary of what the NATO is, we will say that its essential and enduring purpose is to safeguard the freedom and security of all its members by political and military means. Collective defence is at the heart of the Alliance and creates a spirit of solidarity and cohesion among its members.

NATO strives to secure a lasting peace in Europe, based on common values of individual liberty, democracy, human rights and the rule of law. Since the outbreak of crises and conflicts beyond the borders of NATO member countries can jeopardise this objective, the Alliance also contributes to peace and stability through crisis management operations and partnerships. Essentially, NATO not only helps to defend the territory of its members, but engages where possible and when necessary to project its values further afield, prevent crises, manage crises, stabilise post-conflict situations and support reconstruction.

NATO also embodies the transatlantic link by which the security of North America is tied to the security of Europe. It is an intergovernmental organisation which provides a forum where members can consult together on any issues they may choose to raise and make decisions on political and military matters affecting their security. No single member country is forced to rely solely on its national capabilities to meet its essential national security objectives. The resulting sense of shared security among members contributes to stability in the Euro-Atlantic area.

NATO's fundamental security tasks are laid down in the Washington Treaty. They are sufficiently general to withstand the test of time and are translated into more detail in strategic concepts.

Additionally, NATO has carried out a task, not expressly sought in its origins, which is to contribute to peace among European Allies, and of these with the rest of NATO Allies. In this way, the Alliance has been a prop for the maintenance of peace, not only with Russia and the former block of the Warsaw Pact, but also in Western Europe. And we cannot forget the aspect of political and economic cooperation that the Alliance highlights in its Art. 2.

Finally, we should mention that the contents of this book are based mainly on official documents issued by NATO.

# ORIGINS AND FUNDAMENTAL SECURITY TASKS

The North Atlantic Treaty Organization is an alliance of 30 countries from North America and Europe committed to fulfilling the goals of the North Atlantic Treaty signed in Washington on 4 April 1949.

Its original members were Belgium, Canada, Denmark, France, Iceland, Italy, Luxembourg, the Netherlands, Norway, Portugal, the United Kingdom, and the United States. Joining the original signatories were Greece and Turkey (1952); West Germany (1955; from 1990 as Germany); Spain (1982); the Czech Republic, Hungary, and Poland (1999); Bulgaria, Estonia, Latvia, Lithuania, Romania, Slovakia, and Slovenia (2004); Albania and Croatia (2009); Montenegro (2017); and North Macedonia (2020). France withdrew from the integrated military command of NATO in 1966 but remained a member of the organisation; it resumed its position in NATO's military command in 2009.

In accordance with the Treaty, the fundamental role of NATO is to safeguard the freedom and security of its member countries by political and military means. NATO safeguards the Allies' common values of democracy, individual liberty, the rule of law and the peaceful resolution of disputes and promotes these values throughout the Euro-Atlantic area. It provides a forum in which countries from North America and Europe can consult together on security issues of common concern and take joint action in addressing them.

For a long time, relations between North American and European members of the Alliance are the bedrock of

NATO. These countries share the same essential values and interests and are committed to the maintenance of democratic principles, making the security of Europe and that of North America indivisible.

As a basic idea, the Alliance is committed to defending its member states against aggression or the threat of aggression and to the principle that an attack against one or several members would be considered as an attack against all.

It must be clear that NATO remains an inter-governmental organisation in which each member country retains its sovereignty. All NATO decisions are taken jointly by the member countries on the basis of consensus. NATO's most important decision-making body is the North Atlantic Council, which brings together representatives of all the Allies at the level of ambassadors, ministers or heads of state and government. Each member country participates fully in the decision-making process on the basis of equality, irrespective of its size or political, military and economic strength.

Therefore, the Allies retain the scope for independent action regarding joint decisions and actions. However, Allied decisions, once taken, enable unified and concerted action to be reinforced by political solidarity.

This was manifest, for example, in the decisions taken to provide assistance to the United States after the attacks of 11 September 2001. For the first time in its history, NATO invoked Article 5 of the Washington Treaty, which stipulates that an armed attack against one or more members of the Alliance is considered as an armed attack against all. This way, all the members of the Alliance condemned the attacks and provided support to the United States in its response.

It is essential to keep in mind that NATO has no operational forces of its own other than those assigned to it by member countries or contributed by Partner countries for the purpose of carrying out a specific mission. It has a number of mechanisms available to it for this purpose: the defence planning and resource planning processes that form the basis of cooperation within the Alliance, the implementation of political commitments to improved capabilities, and a military structure that combines the functions of a multinational force planning organisation with an Alliance-wide system of command and control of the military forces assigned to it.

In other words, under the command of NATO's strategic commanders, the Organization provides for the joint planning, exercising and operational deployment of forces provided by the member countries in accordance with a commonly agreed force planning process. In sum, an important part of NATO's role is to act as a catalyst for generating the forces needed to meet requirements and enabling member countries to participate in crisis management operations which they could not otherwise undertake on their own.

On the other hand, dialogue and cooperation with non-NATO countries have helped to overcome Cold War era divisions and extend security and stability beyond NATO's borders. The Alliance is deepening and expanding its cooperation with Russia and Ukraine and with other partner countries, some of which have become members, as well as with countries in the Mediterranean Dialogue programme and in the Middle East in general. It is also strengthening cooperation with other international organisations and, in particular, with the European Union, with which it is developing a strategic partnership. NATO structures and mechanisms provide the framework for these various forms of cooperation, which are an integral part of the Alliance's daily activity.

As a summary of what NATO is, we will recall what has been said in the Preface and say that its essential and enduring purpose is to safeguard the freedom and security of all its members by political and military means. Collective defence is at the heart of the Alliance and creates a spirit of solidarity and cohesion among its members.

NATO strives to secure a lasting peace in Europe, based on common values of individual liberty, democracy, human rights and the rule of law. Since the outbreak of crises and conflicts beyond the borders of NATO member countries can jeopardise this objective, the Alliance also contributes to peace and stability through crisis management operations and partnerships.

Essentially, NATO not only helps to defend the territory of its members, but engages where possible and when necessary to project its values further afield, prevent crises, manage crises, stabilise post-conflict situations and support reconstruction.

NATO also embodies the transatlantic link by which the security of North America is tied to the security of Europe. It is an intergovernmental organisation which provides a forum where members can consult together on any issues they may choose to raise and make decisions on political and military matters affecting their security. No single member country is forced to rely solely on its national capabilities to meet its essential national security objectives. The resulting sense of shared security among members contributes to stability in the Euro-Atlantic area.

NATO's fundamental security tasks are laid down in the Washington Treaty. They are sufficiently general to withstand the test of time and are translated into more detail in strategic concepts. Strategic concepts are the authoritative statement of the Alliance's objectives and provide the highest level of guidance on the political and military means to be used in achieving these goals; they remain the basis for the implementation of Alliance policy as a whole.

During the Cold War, NATO focused on collective defence and the protection of its members from potential threats emanating from the Soviet Union. With the collapse of the Soviet Union, along with the rise of non-state actors affecting international security, many new security threats emerged. NATO now focuses on countering these threats by utilising collective defence, managing crisis situations and encouraging cooperative security, as outlined in the 2010 Strategic Concept.

Additionally, NATO has carried out a task, not expressly sought in its origins, which is to contribute to peace among European Allies, and of these with the rest of NATO Allies. In this way, the Alliance has been a prop for the maintenance of peace, not only with Russia and the former block of the Warsaw Pact, but also in Western Europe. And we cannot forget the aspect of political and economic cooperation that the Alliance highlights in its Art. 2.

**The origins of the Alliance**

Let us recall first that from 1945 to 1949, faced with the pressing need for economic reconstruction, the countries of Western Europe and their North American allies viewed with concern the expansionist policies and methods of the USSR. Having fulfilled their own postwar commitments to shrink their defence establishments and demobilise their forces, Western governments became increasingly alarmed when it became clear that the Soviet leadership intended to maintain its own military forces in full force.

Furthermore, in view of the declared ideological objectives of the Soviet Communist Party, it was evident that calls for respect for the Charter of the United Nations, and for respect for the international settlements reached at the end of the Second World War, would not guarantee the national sovereignty or independence of democratic states in the face of the threat of external aggression or internal subversion. The imposition of undemocratic forms of government and the repression of effective opposition and basic human and civil rights and freedoms in many countries in Central and Eastern Europe, as well as in other parts of the world, compounded these fears.

Thus, between 1947 and 1949, a series of dramatic political events put things at a critical point. These included direct threats to the sovereignty of Norway, Greece, Turkey and other Western European countries, the June 1948 coup in Czechoslovakia and the illegal blockade of Berlin that began in April of the same year. The Brussels Treaty was signed by the United Kingdom, France, Belgium, Luxembourg and the Netherlands on 17 March 1948. This Treaty established the Western Union (WU) which was an intergovernmental defence alliance that also promoted economic, cultural and social collaboration.

The signing of the Brussels Treaty marked the determination of five Western European to develop a common defence system and strengthen ties between them in a way that would allow them to resist ideological, political and military threats to their security.

The Brussels Treaty represented the first step in the post-war reconstruction of western european security and brought the Western Union Defence Organisation (the defence arm of the WU) into being. It was also the first step in the process leading to the signature of the North Atlantic Treaty in 1949 and the creation of the North Atlantic Alliance.

Negotiations with the United States and Canada then followed on the creation of a single North Atlantic Alliance based on security guarantees and mutual commitments between Europe and North America. Denmark, Iceland, Italy, Norway and Portugal were invited by the Brussels Treaty powers to become participants in this process. These negotiations culminated in the signature of the Washington Treaty in April 1949, which introduced a common security system based on a partnership among these 12 countries.

The North Atlantic Alliance was founded on the basis of a treaty between member states entered into freely by each of them after public debate and due parliamentary process. The Treaty upholds their individual rights as well as their international obligations in accordance with the Charter of the United Nations. Through the treaty, member countries commit themselves to sharing the risks and responsibilities of collective security and undertake not to enter into any other international commitments which might conflict with the treaty.

Thus, since NATO's creation, its central focus has been to provide for the immediate defence and security of its member countries. Today this remains its core task, but its main focus has undergone fundamental changes to enable the Alliance to confront new threats and meet new challenges.

## NATO's key security tasks

The Washington Treaty says that NATO's essential and enduring purpose is to safeguard the freedom and security of all its members by political and military means. Based on common values of democracy, human rights and the rule of law, the Alliance has been striving since its inception to ensure a lasting peaceful order in Europe.

However, the achievement of this objective may be compromised by crises and conflicts outside the Euro-Atlantic area. Therefore, the Alliance not only guarantees the defence of its members, but contributes to peace and stability beyond the geographic space defined as the North Atlantic Treaty area through associations and crisis management operations.

More specifically, the guiding principle by which the Alliance works is common commitment and mutual cooperation between sovereign states in support of the indivisibility of security for all its members. Solidarity and cohesion within the Alliance, through daily cooperation in the political and military spheres, ensure that no member country is forced to rely solely on its own national efforts to meet basic security challenges.

Without depriving member countries of their right and duty to assume their sovereign responsibilities in the field of defence, the Alliance allows them through collective efforts to meet their essential national security objectives.

Core security tasks of the Allliance are outlined in the NATO's Strategic Concept. It is the authoritative statement of the objectives of the Alliance and provides the highest level of guidance on the political and military means that will be used to achieve them. It remains the basis for the implementation of the Alliance's policy as a whole. However, changing threats and perceptions of threats have resulted in an ongoing process of adapting this strategy to ensure that the political framework, military structures and military capabilities necessary to meet modern security challenges are in place.

First published in 1991, the Strategic Concept differed dramatically from preceding documents both in content and form. It maintained the security of its members as NATO's fundamental purpose but combined this with the specific obligation to work towards improved and expanded security for Europe as a whole through partnership and cooperation with former adversaries. In addition, it was issued as a public document, open for discussion and comment by parliaments, security specialists, journalists and the broader public. The Strategic Concept was revised in 1999 and 2010, committing the Allies not only to common defence but to the peace and stability of the wider Euro-Atlantic area. It comprises the following political elements:

- It reconfirms the bond between our nations to defend one another against attack, including against new threats to the safety of our citizens.

- It commits the Alliance to prevent crises, manage conflicts and stabilise post-conflict situations, including by working more closely with our international partners, most importantly the United Nations and the European Union.

- It offers our partners around the globe more political engagement with the Alliance, and a substantial role in shaping the NATO-led operations to which they contribute.

- It commits NATO to the goal of creating the conditions for a world without nuclear weapons; but reconfirms that, as long as there are nuclear weapons in the world, NATO will remain a nuclear Alliance.

- It restates our firm commitment to keep the door to NATO open to all European democracies that meet the standards of membership, because enlargement contributes to our goal of a Europe whole, free and at peace.

- It commits NATO to continuous reform towards a more effective, efficient and flexible Alliance, so that our taxpayers get the most security for the money they invest in defence.

The 2010 Strategic Concept "Active Engagement, Modern Defence" is a very clear and resolute statement on NATO's core tasks and principles, its values, the evolving security environment and the Alliance's strategic objectives for the next decade.

After having described NATO as "a unique community of values committed to the principles of individual liberty, democracy, human rights and the rule of law", it presents NATO's three essential core tasks: collective defence, crisis management and cooperative security. It also emphasises Alliance solidarity, the importance of transatlantic consultation and the need to engage in a continuous process of reform.

The document then describes the current security environment and identifies the capabilities and policies it will put into place to ensure that NATO's defence and deterrence, as well as crisis management abilities, are sufficiently well equipped to face today's threats.

These threats include, for instance, the proliferation of ballistic missiles and nuclear weapons, terrorism, cyber attacks and fundamental environmental problems. The Strategic Concept also affirms how NATO aims to promote international security through cooperation. It will do this by reinforcing arms control, disarmament and non-proliferation efforts, emphasising NATO's open door policy for all European countries, and significantly enhancing its partnerships in the broad sense of the term. Additionally, it affirms that NATO will continue its reform and transformation process.

NATO's essential core tasks and principles:

After having reiterated NATO's enduring purpose and key values and principles, the Strategic Concept highlights the Organization's core tasks.

The modern security environment contains a broad and evolving set of challenges to the security of NATO's territory and populations. In order to assure their security, the Alliance must and will continue fulfilling effectively three essential core tasks, all of which contribute to safeguarding Alliance members, and always in accordance with international law:

1. Collective defence: NATO members will always assist each other against attack, in accordance with Article 5 of the Washington Treaty. That commitment remains firm and binding. NATO will deter and defend against any threat of aggression, and against emerging security challenges where they threaten the fundamental security of individual Allies or the Alliance as a whole.

2. Crisis management: NATO has a unique and robust set of political and military capabilities to address the full spectrum of crises, before, during and after conflicts. NATO will actively employ an appropriate mix of those political and military tools to help manage developing crises that have the potential to affect Alliance security, before they escalate into conflicts; to stop ongoing conflicts where they affect Alliance security; and to help consolidate stability in post-conflict situations where that contributes to Euro-Atlantic security.

3. Cooperative security: The Alliance is affected by, and can affect, political and security developments beyond its borders. The Alliance will actively engage to enhance

international security, through partnership with relevant countries and other international organisations; by contributing actively to arms control, non-proliferation and disarmament; and by keeping the door of membership in the Alliance open to all European democracies that meet NATO's standards.

Deterrence and defence:

The 2010 Strategic Concept states that collective defence is the Alliance's greatest responsibility and "deterrence, based on an appropriate mix of nuclear and conventional capabilities, remains a core element" of NATO's overall strategy. While stressing that the Alliance does not consider any country to be its adversary, it provides a comprehensive list of capabilities the Alliance aims to maintain and develop to counter existing and emerging threats. These threats include the proliferation of nuclear weapons, ballistic missiles and other weapons of mass destruction and their means of delivery; terrorism, cyber-attacks and key environmental and resource constraints.

Crisis management:

NATO is adopting a holistic approach to crisis management, envisaging NATO involvement at all stages of a crisis: "NATO will therefore engage, where possible and when necessary, to prevent crises, manage crises, stabilise post-conflict situations and support reconstruction". It is encouraging a greater number of actors to participate and coordinate their efforts and is considering a broader range of tools to be more effective across the crisis management spectrum. This comprehensive, all-encompassing approach to crises, together with greater emphasis on training and developing local forces goes hand-in-hand with efforts to enhance civil-military planning and interaction.

Cooperative security:

The final part of the 2010 Strategic Concept focuses on promoting international security through cooperation. At the root of this cooperation is the principle of seeking security "at the lowest possible level of forces" by supporting arms control, disarmament and non-proliferation. NATO states that it will continue to help reinforce efforts in these areas and cites a number of related initiatives. It then recommits to NATO enlargement as the best way of achieving "our goal of a Europe whole and free, and sharing common values".

A fundamental component of its cooperative approach to security is partnership, understood between NATO and non-NATO countries, as well as with other international organisations and actors.

The Strategic Concept depicts a more inclusive, flexible and open relationship with the Alliance's partners across the globe and stresses its desire to strengthen cooperation with the United Nations and the European Union. It also reiterates its commitment to developing relations with countries of the Mediterranean and the Gulf region.

Finally, the Strategic Concept describes the means NATO will use to maximise efficiency, improve working methods and spend its resources more wisely in view of the priorities identified in this concept.

**Facing the changing security environment**

Let us point out that the historic decision taken by NATO to invoke Article 5 of the Washington Treaty and extend its assistance to the United States following 11 September 2001 marked the beginning of a new impetus in NATO's transformation process that was to touch on virtually every aspect of Alliance activity.

In addition to combating terrorism, a variety of other factors have reinforced the need for adaptation of Alliance structures and policies. These include the increased threat posed by weapons of mass destruction and the need for new operational capabilities in critical areas. The demands of NATO's enlargement have also had an impact, as have the developing role of partnerships with Russia, Ukraine and partner countries, the importance of the Mediterranean Dialogue and the Istanbul Cooperation Initiative, and the strategic partnership with the European Union. NATO's leading role in the International Security Assistance Force in Afghanistan and its continuing role in the Balkans have also led the Organization to adapt itself to the requirements of these operations, of its missions in Iraq and Sudan, and of its relief efforts in Pakistan.

Many of the changes needed to carry forward the transformation process were introduced at NATO's Prague Summit on 21–22 November 2002 and were pursued at its Istanbul Summit on 28–29 June 2004. Five major areas have been affected: membership of the Alliance, the reform of NATO's civilian and military structures, the acceptance of new roles, the development of new capabilities and the promotion of new relationships.

In December 2019, NATO Leaders invited the Secretary General to lead a forward-looking reflection process to strengthen NATO's political dimension. Thus, on 8 June 2020, NATO Secretary General Jens Stoltenberg launched his outline for NATO 2030. "This is an opportunity to reflect on where we see our Alliance ten years from now, and how it will continue to keep us safe in a more uncertain world" the Secretary General said.

To do this, NATO must "stay strong militarily, be more united politically, and take a broader approach globally". Mr. Stoltenberg stated that: a) staying strong militarily means continuing to invest in the armed forces and modern military capabilities; b) strengthening NATO politically means using NATO as the forum to discuss, and where necessary to act, on issues affecting the shared security; c) finally, making NATO a more global Alliance means working even more closely with like-minded partners to defend values in a world of increased global competition.

# POLICY AND DECISION-MAKING IN NATO

The fundamental rule is that the principle of consensus decision-making is applied throughout the Alliance, reflecting the fact that it is the member countries that decide and each one of them is involved in the decision-making process. This principle is applied at every level of the Organization.

The principal policy and decision-making institutions of the Alliance are the North Atlantic Council, the Defence Planning Committee and the Nuclear Planning Group. Each of these plays a vital role in the consultative and decision-making processes that are the bedrock of the cooperation, joint planning and shared security between member countries that NATO represents.

It is important to know that the decisions taken by each of these bodies have the same status and represent the agreed policy of the member countries, irrespective of the level at which they are taken. Subordinate to these senior bodies are specialised committees also consisting of officials representing their countries. This committee structure provides the basic mechanism that gives the Alliance its consultation and decision-making capability, ensuring that each member country can be represented at every level and in all fields of NATO activity.

To be clearer, NATO decisions are taken on the basis of consensus, after discussion and consultation among member countries. A decision reached by consensus is an agreement reached by common consent and supported by each member country. This implies that when a NATO decision is taken, it is the expression of the collective will of the sovereign states that are members of the Alliance. It is this decision-making process that gives NATO both its strength and its credibility.

When there is disagreement, discussions take place until a decision is reached, and in some circumstances this may be to recognise that agreement is not possible. In general, however, mutually acceptable solutions are normally found. The process is rapid since members consult on a continuous basis and therefore frequently know and understand each other's positions in advance. Consultation is a vital part of the decision-making process. It facilitates communication between members whose prime goal is to ensure that decisions taken collectively are consistent with their national interests.


**The North Atlantic Council**

The North Atlantic Council (NAC) has effective political authority and powers of decision, and consists of permanent representatives of all member countries meeting together at least once a week. The Council also meets at higher levels involving foreign ministers, defence ministers or heads of state and government, but it has the same authority and powers of decision-making, and its decisions have the same status and validity, at whatever level it meets.

The Council has an important public profile and issues declarations and communiqués explaining the Alliance's policies and decisions to the general public and to governments of countries which are not members of NATO.

Being the most important organ of NATO, the Council is the only body within the Alliance which derives its authority explicitly from the North Atlantic Treaty. The Council itself was given responsibility under the Treaty for setting up subsidiary bodies. Many committees and planning groups have since been created to support the work of the Council or to assume responsibility in specific fields such as defence planning, nuclear planning and military matters.

Thus, the Council provides a unique forum for wide-ranging consultation between member governments on all issues affecting their security and is the most important decision-making body in NATO. All member countries of NATO have an equal right to express their views round the Council table. Decisions are the expression of the collective will of member governments arrived at by common consent. All member governments are party to the policies formulated in the Council or under its authority and share in the consensus on which decisions are based.

When the Council meets at the level of ambassadors or permanent representatives of the member countries, it is often referred to as the "Permanent Council". Twice a year, and sometimes more frequently, it meets at ministerial level, either in formal or informal session, when each country is represented by its minister of foreign affairs. Meetings of the Council also take place in defence ministers' sessions. Summit meetings attended by heads of state or government are held whenever particularly important issues have to be addressed or at seminal moments in the evolution of Allied security policy.

Besides, while the Council normally meets at least once a week, it can be convened at short notice whenever necessary. Its meetings are chaired by the Secretary General of NATO or, in his absence, by his Deputy. The longest serving permanent representative on the Council assumes the title of Dean of the Council. Primarily a ceremonial function, the Dean may be called upon to play a more specific presiding role, for example in convening meetings and chairing discussions at the time of the selection of a new secretary general.

At ministerial meetings of foreign ministers, one country's foreign minister assumes the role of honorary president. The position rotates annually among the member countries in the order of the English alphabet. An order of precedence in the Permanent Council is established on the basis of length of service, but at meetings of the Council at any level, permanent representatives sit round the table in order of nationality, in English alphabetical order. The same procedure is followed throughout the NATO committee structure.

In NATO items discussed and decisions taken at meetings of the Council cover all aspects of the Organization's activities and are frequently based on reports and recommendations prepared by subordinate committees at the Council's request.

Equally, subjects may be raised by any one of the national representatives or by the Secretary General. Permanent representatives act on instructions from their capitals, informing and explaining the views and policy decisions of their governments to their colleagues round the table. Conversely they report back to their national authorities on the views expressed and positions taken by other governments, informing them of new developments and keeping them abreast of movement towards consensus on important issues or areas where national positions diverge.

It is important to note that when decisions have to be taken, action is agreed upon on the basis of unanimity and common accord. There is no voting or decision by majority. Each member country represented at the Council table or on any of its subordinate committees retains complete sovereignty and responsibility for its own decisions.

The work of the Council is prepared by subordinate Committees with responsibility for specific areas of policy. Much of this work involves the Senior Political Committee (SPC), consisting of deputy permanent representatives, sometimes reinforced by appropriate national experts, depending on the subject. In such cases it is known as the SPC(R). The Senior Political Committee has particular responsibility for preparing most statements or communiqués to be issued by the Council and meets in advance of ministerial meetings to draft such texts for Council approval. Other aspects of political work may be handled by the regular Political Committee, which consists of political counsellors or advisers from national delegations. Similarly, the work of the Defence Planning Committee and the Nuclear Planning Group (NPG) is prepared by the Defence Review Committee and the NPG Staff Group respectively, and by other senior committees.

On the other hand, when the Council meets at the level of defence ministers or is dealing with defence matters and questions relating to defence strategy, other senior committees, such as the Executive Working Group, may be involved as the principal advisory bodies. If financial matters are on the Council's agenda, the Senior Resource

Board, or the Civil or Military Budget Committees, or the Infrastructure Committee, depending on which body is appropriate, will be responsible to the Council for preparing its work. Depending on the topic under discussion, the respective senior committee with responsibility for the subject area assumes the leading role in preparing Council meetings and following up on Council decisions.

And of course, the work of the Council is supported by the relevant divisions and offices of the International Staff, and in particular by the Council Secretariat, which coordinates Council activities and ensures that Council mandates are executed and its decisions recorded and disseminated.


**The Defence Planning Committee**

Due to its importance in the Atlantic structure, the Defence Planning Committee (DPC) is normally composed of permanent representatives but meets at the level of defence ministers at least twice a year, and deals with most defence matters and subjects related to collective defence planning. The Defence Planning Committee provides guidance to NATO's military authorities and, within its scope of activity, has the same functions and attributes and the same authority as the Council on matters within its area of responsibility.

The work of the Defence Planning Committee is prepared by a number of subordinate committees with specific responsibilities and in particular by the Defence Review Committee, which oversees the force planning process within NATO and examines other issues relating to the integrated military structure. Like the Council, the Defence

Planning Committee looks to the senior committee with the relevant specific responsibility for the preparatory and follow-up work arising from its decisions.


**The Nuclear Planning Group**

Equally, the Defence Ministers of member countries which take part in NATO's Defence Planning Committee meet at regular intervals in the Nuclear Planning Group (NPG), where they discuss specific policy issues associated with nuclear forces. These discussions cover a broad range of nuclear policy matters including the safety, security and survivability of nuclear weapons, communications and information systems, deployment issues and wider questions of common concern such as nuclear arms control and nuclear proliferation. The Alliance's nuclear policy is kept under review and decisions are taken jointly to modify or adapt it in the light of new developments and to update and adjust planning and consultation procedures.

Habitually, the work of the Nuclear Planning Group is prepared by an NPG Staff Group composed of members of the national delegations of the countries participating in the NPG, members of the International Military Staff and representatives of the Strategic Commanders. The Staff Group carries out detailed work on behalf of the NPG Permanent Representatives. It meets once a week and at other times as necessary.

It should also be mentioned that the High Level Group (HLG) is a senior advisory body to the NPG on nuclear policy and planning issues. The High Level Group is also charged with overseeing nuclear weapons safety, security and survivability matters. The Group is chaired by the United States and is composed of national policymakers and experts from capitals as well as members of NATO's International Staffs and representatives of the Strategic Commanders. It meets several times a year to discuss aspects of NATO's nuclear policy, planning and force posture, and matters concerning the safety, security and survivability of nuclear weapons.


## The Military Committee

And thirdly, the Military Committee is the senior military authority in NATO under the overall political authority of the Council, the Defence Planning Committee or the Nuclear Planning Group. It is an integral part of the policy and decision-making apparatus of the Alliance and provides an essential link between the political decision-making process within the North Atlantic Council, Defence Planning Committee and Nuclear Planning Group and the integrated command structures of NATO charged respectively with the conduct of military operations and the further military transformation of the Alliance.

Also the Military Committee is responsible for overseeing the development of NATO's military policy and doctrine and for providing guidance to the NATO Strategic Commanders. The Strategic Commanders are responsible to the Military Committee for the overall direction and conduct of all Alliance military matters within their fields of responsibility. The Military Committee is supported in its activities by the International Military Staff.

**The consultative process**

It is essential to keep in mind that policy formulation and implementation, in an Alliance of independent sovereign countries, depends on all member governments being fully informed of each other's overall policies and intentions as well as the underlying considerations which give rise to them. This calls for regular political consultation, whenever possible during the policy-making stage of deliberations before national decisions have been taken.

Thus, political consultation in NATO began as a systematic exercise when the Council first met in September 1949, shortly after the North Atlantic Treaty came into force. Since that time it has been strengthened and adapted to suit new developments. The principal forum for political consultation remains the Council. Its meetings take place with a minimum of formality; discussion is frank and direct. The Secretary General, by virtue of his chairmanship, plays an essential part in its deliberations and acts as its principal representative and spokesman both in contacts with individual governments and in public affairs.

Consultation also takes place on a regular basis in other fora, all of which derive their authority from the Council. The Political Committee at senior and other levels, the Policy Coordination Group, the Atlantic Policy Advisory Group and other special committees all have a direct role to play in facilitating political consultation between member governments. Like the Council, they are assisted by an International Staff responsible to the Secretary General of NATO.

Political consultation among the members of the Alliance is not limited to events taking place within the Euro-Atlantic area. Events elsewhere that have potential implications for the Alliance regularly feature on the agenda of the Council and its subordinate committees. The consultative machinery of NATO is readily available and extensively used by the members in such circumstances, in order to identify at an early stage areas where, in the interests of security and stability, coordinated action may be taken.

Neither is the need for consultation limited to political subjects. Wide-ranging consultation takes place in many other fields. The process is continuous and takes place on an informal as well as a formal basis with a minimum of delay or inconvenience, as a result of the collocation of national delegations to NATO within the same headquarters. Where necessary, it enables intensive work to be carried out at short notice on matters of particular importance or urgency with the full participation of representatives from all the governments concerned.

In its most basic form it simply involves the exchange of information and opinions. But consultation within the Alliance takes many forms. At another level, it covers the communication of actions or decisions which governments have already taken or may be about to take and which have a direct or indirect bearing on the interests of their Allies.

It may also involve providing advance warning of actions or decisions to be taken by governments in the future, in order to provide an opportunity for them to be endorsed or commented upon by others. It can encompass discussion with the aim of reaching a consensus on policies to be adopted or actions to be taken in parallel. And ultimately it is designed to enable member countries to arrive at mutually acceptable agreements on collective decisions or on action by the Alliance as a whole.

With this particular underlying philosophy, by making their joint decision-making process dependent on consensus and common consent, the members of the Alliance safeguard the role of each country's experience and individual perspectives, while taking advantage of the machinery and the procedures that allow them to act jointly, quickly and decisively if the circumstances require it.

The practice of exchanging information and consulting together on a daily basis ensures that governments can come together at short notice whenever necessary, often with prior knowledge of their respective preoccupations, in order to agree on common policies. If need be, efforts to reconcile differences between them will be made in order that joint actions may be backed by the full force of decisions to which all the member governments subscribe. Once taken, such decisions represent the common determination of all the countries involved to implement them in full. Decisions which may be politically difficult, or which face competing demands on resources, thus acquire added force and credibility.

We must repeat that all NATO member countries participate in the political level of cooperation within the Alliance and are equally committed to the terms of the North Atlantic Treaty, in particular to the reciprocal commitment made in Article 5 that symbolises the indivisibility of their security, that is, consider an attack against one or more of them as an attack against all of them.

Furthermore, the way the Alliance has evolved ensures that variations in the requirements and policies of member countries can be taken into account in their positions within the Alliance. This flexibility manifests itself in several different ways. In some cases, the differences can be largely procedural and are accommodated without difficulty. Iceland, for example, has no military forces and is therefore represented at NATO military forums by a civilian if it so desires. In other cases, the distinctions may be more substantive in nature. France, a founding member of the Alliance in 1949, withdrew from the Alliance's integrated military structure in 1966 while remaining a full member of its political structures.

Ultimately, distinctions between NATO member countries may also exist as a result of their geographical, political, military or constitutional situations. Norway and Denmark's participation in NATO military provisions, for example, must comply with national legislation that does not allow foreign forces or nuclear weapons to be stationed on their national territory in peacetime. In another context, military commands within the integrated military structure may involve only forces from those countries directly interested or equipped to participate in the specific function for which the command has been created.

**Consultations with Partner countries, other non-member countries and contact countries**

Another important feature of the Alliance is that cooperation with non-NATO countries is an integral part of the Alliance's security policy and plays a fundamental role in its daily work. By seeking cooperation and different forms of association with non-member countries, NATO not only increases security and stability for its partner countries, but also strengthens its own security. Partnership and cooperation are therefore part of a two-way process that benefits both partner countries and member countries. It provides an opportunity for each of them to discuss security issues and cooperate in different fields, helping to overcome divisions and possible areas of disagreement that could lead to instability and conflict.

This way of doing is materialised by holding regular consultations on relevant political issues with partner countries in the context of the Euro-Atlantic Partnership Council, with Russia through the NATO-Russia Council, with Ukraine through the NATO-Ukraine Commission, and with participants in NATO's Mediterranean Dialogue through the Mediterranean Cooperation Group. NATO has also offered a framework for cooperation with countries of the broader Middle East, through the Istanbul Cooperation Initiative, and maintains a consultative forum for cooperation with countries in the Balkans, through the South-East Europe Initiative.

The guidelines governing consultations in these forums are based on those which for a long time formed the basis for consultations within the Alliance itself and are carried out with the same openness and spirit of cooperation. Finally, there are provisions for NATO consultations with any active participant in the Partnership for Peace, if that Partner perceives a direct threat to their territorial integrity, political independence or security.

Thus, the process of cooperation at the national level is reinforced by cooperation between NATO and a number of other multinational organisations with a critical role to play in security-related matters. NATO does not therefore work in isolation. In addition to the tasks in which it plays the leading role, it acts to support and complement the work of other organisations in laying the foundation for a safer, more stable and more peaceful international environment in which economies can prosper and individuals flourish.

In particular, NATO has undertaken military operations to support the principles and resolutions of the United Nations. It is working closely with its European member countries in developing an effective strategic partnership between the Alliance and the European Union. And the Alliance works closely in different contexts with the Organization for Security and Co-operation in Europe (OSCE), the Council of Europe and other international organisations and non-governmental organisations.

On the other hand, despite NATO has no formal institutional links with individual countries outside the framework of the bilateral and multilateral structures described above, the Alliance's role in the security of today's world leads many other countries to seek up-to-date information about NATO policies and activities, to remain in touch and to consider participating in specific projects. The various operational roles undertaken by the Alliance have also served to increase interaction with countries contributing to such efforts.

In such cases, in accordance with guidance issued by the North Atlantic Council, cooperation is considered on a case-by-case basis. Decisions are taken in the light of mutual benefits, potential costs, the priority given to cooperation with Partner countries and the extent to which the values that the Alliance represents are shared.

Besides, contacts and exchanges take place with various countries, called "contact countries", which have expressed their desire to establish a dialogue with the Alliance. For several years, NATO has engaged in a regular exchange of views at all levels with Japan. More recently, the Alliance has also responded positively to China's interest in informal contacts. Regular contacts have also been developed at all levels with other countries such as New Zealand and Australia. In some cases, these dialogues can be supplemented by participation in specific NATO activities or joint participation in events.

To make this concrete, we will say that NATO-led operations in the Balkans, the Mediterranean and Afghanistan, as well as the training mission in Iraq agreed in 2004, provide concrete examples of practical cooperation between the Alliance and countries that do not they are not members of or linked to it through formal associations. Countries that have contributed forces to these operations include Argentina, Australia, Chile, New Zealand and the United Arab Emirates.


**Crisis management**

The Alliance's Strategic Concept defines crisis management as one of NATO's fundamental security tasks. It commits the Alliance to be prepared to contribute to effective conflict prevention and to actively participate in crisis management, including crisis response operations. This requirement is met through a combination of effective consultation procedures, crisis management arrangements, military capabilities, and civil emergency planning preparations.

From the first years of its existence, NATO's basic task was to develop a defence planning process combined with the necessary military capabilities to ensure that the Alliance had the capacity to manage collective defence operations under Article 5 of the Treaty. However, it is only during its most recent history that NATO has made decisions that have resulted in operations outside Article 5 outside the territory of member countries, designed to prevent a conflict from spreading and threatening to destabilise other countries of the region, including NATO Members or partner countries. Simultaneously, the Alliance has undertaken a series of measures to develop its capacity to respond to non-Article 5 crisis situations.

This way, an increasingly important part of NATO's role in the years since the end of the Cold War has therefore been the unique contribution it has been able to make to efforts by the wider international community to prevent conflict from occurring and, when it does occur, to restore and preserve peace. In June 1992, at a meeting of the North Atlantic Council in Oslo, NATO offered to support, on a case-by-case basis and in accordance with its own procedures, peacekeeping and other operations under the responsibility of the Conference for Security and Co-operation in Europe (which became the Organization for Security and Co-operation in Europe (OSCE) in 1995). A few months later, the same commitment was made with respect to peacekeeping operations under the authority of the United Nations. The Alliance stood ready to respond positively to initiatives that the United Nations Secretary

General might take in seeking NATO assistance in implementing United Nations Security Council resolutions.

Later, between 1992 and 1995 NATO undertook a number of monitoring and enforcement operations in support of successive UN Security Council Resolutions relating to the continuing crisis and deteriorating situation in the former Yugoslavia. However, the NATO-led Implementation Force (IFOR) in Bosnia and Herzegovina, created in December 1995 to implement the military aspects of the Dayton Peace Agreement that ended the Bosnian conflict, was the first major manifestation of this policy. Since that time the Alliance has undertaken further peace-support operations and crisis management tasks in Bosnia and Herzegovina, Kosovo, the former Yugoslav Republic of Macedonia and Afghanistan, in cooperation with the United Nations, the OSCE and the European Union. In August 2003 it took over the leadership of the International Security Assistance Force (ISAF) in Afghanistan. In August 2004, a NATO Training Mission for Iraq was established to assist the Iraqi government in training and building up its own national security forces.

But as a fundamental concept in this matter we will say that crisis management can involve both military and non-military measures to respond to a crisis situation threatening national or international security. A crisis may be essentially political in nature, or military, or humanitarian, and may be caused by political disputes or armed conflict, technological incidents or natural disasters. Crisis management consists of the different means of dealing with these varying forms of crises.

The national or international response to a crisis, or to an evolving situation that threatens to become a crisis, depends on the nature, scale and seriousness of the situation. In some cases, it may be possible to anticipate and prevent a crisis through diplomacy or other measures. At other times more robust measures may be necessary, including military action. Moreover, depending on the nature of the crisis, different types of crisis management operation may be contemplated by national authorities.

Nowadays there are two broad categories of crisis management operations within NATO that member countries may consider, namely operations calling for collective defence, and other crisis response operations in which collective defence is not involved:

- First, collective defence operations are based on the invocation of Article 5 of the North Atlantic Treaty and are referred to as "Article 5 operations". They carry the implication that the decision has been taken collectively by NATO members to consider an attack or act of aggression against one or more members as an attack against all. As we already know, NATO has invoked Article 5 once in its history, in September 2001, following the terrorist attacks against the United States.

  It is essential to know that readiness to respond to a crisis threatening the security of its member countries by invoking Article 5 of the Treaty and by implementing the mutual guarantees called for under Article 5 has been a fundamental obligation of NATO member countries from the outset. As such it plays an integral part in NATO's defence planning arrangements, which are designed to deter any possible threat and to stand ready, should deterrence fail, to take the action decided upon by the member countries at the political level to restore and maintain the security of the North Atlantic area. Throughout the Cold War years it was widely assumed that the only circumstances in which Article 5 would have to be invoked would be a crisis threatening the security of the European Allies.

- Secondly, other crisis response operations include all military operations that the Alliance may decide to conduct in a non-Article 5 situation. They may be designed to support the peace process in a conflict area and, in those circumstances, are referred to as peace support operations. However, they include a range of other possibilities including conflict prevention, peacekeeping and peace enforcement measures, peace-making, peace-building, preventive deployment and humanitarian operations. Thus, NATO's involvement in the Balkans and Afghanistan are examples of crisis management operations in this category. Other illustrations include NATO's supporting role for Polish troops participating in the International Stabilisation Force in Iraq and the acceptance of responsibility for assisting the Iraqi government with the training of its national security forces by launching the NATO Training Mission for Iraq referred to above.

Also natural, technological or humanitarian disasters can lead to interventions that fall under the category of crisis response operations and involve operations to assist member and partner countries that are victims of major incidents. NATO provided assistance to Pakistan after the catastrophic earthquake in October 2005 and, on different occasions, also assisted Ukraine, which has often been devastated by floods.

The need for the Alliance to consider undertaking military operations in response to non-Article 5 situations emerged during the early years following the end of the Cold War, the disintegration of the Warsaw Pact and the collapse of the Soviet Union. In a number of areas both within and on the borders of the former Soviet Union and in the Balkans, past tensions resurfaced and violent conflicts broke out among ethnic groups, whose rights in many cases had been suppressed for half a century. When major ethnic conflict broke out in the former Yugoslavia in 1992 and repeated international efforts failed to resolve the crisis, NATO member governments took a series of unprecedented decisions to use the Alliance's military capabilities in an operational role.

In all these situations NATO decides whether to engage in a crisis management operation on a case-by-case basis. Such decisions, as with all other Alliance decisions, are based on a consensus among the member countries. In many of the operational situations in which it has taken on responsibilities, cooperation and partnership with other organisations has been an important factor. Effective cooperation with the United Nations and with United Nations agencies on the ground, cooperation with the Organization for Security and Co-operation in

Europe (OSCE), NATO's growing strategic partnership with the European Union in which support has been made available for EU-led operations using NATO assets and capabilities: all these have played a significant role in meeting the specific needs of different forms of crisis. Equally significant has been the Alliance's expanding cooperation in crisis management situations with non-NATO countries that are partners in the Euro-Atlantic Partnership Council (EAPC) or in the Alliance's Mediterranean Dialogue.

At this point we have to refer to the idea that crisis management is a broad concept that goes beyond military operations and can include issues such as protecting populations under threat or victims of natural or man-made disasters. NATO began developing civil protection measures for the eventuality of a nuclear conflict as early as the 1950s and was able to take advantage of capabilities in this field to mitigate the effects of disasters caused by major floods, earthquakes, incidents involving large industrial or technological disasters, and humanitarian crises.

Histologically, it was in 1953 when the first disaster assistance scheme was implemented after devastating floods in northern Europe, and in 1958 NATO established detailed procedures for coordinating assistance between member countries in the event of a disaster. These procedures remained in place and provided the basis for civil emergency planning work in subsequent years. They were comprehensively reviewed in 1995 when they became applicable to Partner countries in addition to NATO member countries.

Similarly, in 1998 a Euro-Atlantic Disaster Response Co-ordination Centre (EADRCC) was established, on the basis of a Russian proposal, to coordinate aid provided by different member and Partner countries to a disaster-stricken area in any one of them. A Euro-Atlantic Disaster Response Unit was also established based on non-permanent civil and military elements volunteered by member or Partner countries for deployment to a disaster area. Soon after its creation, the EADRCC was called upon to help to coordinate humanitarian assistance for Kosovo refugees in support of the United Nations High Commissioner for Refugees. Civil emergency planning measures have also enabled intervention in numerous civil emergencies in cases of flooding in Albania, Czech Republic, Hungary, Romania, Ukraine and the United States; earthquakes in Turkey and Pakistan; fires in the former Yugoslav Republic of Macedonia and in Portugal; and extreme weather conditions knocking out the power supply in Ukraine and Moldova. NATO also conducts civil emergency planning exercises on a regular basis.

**Decision-making on crisis management**

Regarding decision-making on crisis management arrangements, decisions are taken by the governments of NATO member countries collectively in the framework of the North Atlantic Council. They may include political and military measures as well as measures to deal with civil emergencies, depending on the nature of the crisis.

Therefore, all decisions on the planning, deployment or use of military forces are taken only with the political authorisation of the member countries. Such decisions may result in the use of different mechanisms to deal with the crises such as exchanging intelligence, information and other data, comparing different perceptions and approaches, and other measures aimed at harmonising views among the member countries.

In reaching and implementing its decisions, the Council may be supported by specialised committees such as the Political Committee, the Policy Coordination Group, the Military Committee and the Senior Civil Emergency Planning Committee. It will also make full use of and draw on the communications and information systems available to it, including the NATO Situation Centre, which collects and disseminates political, economic and military intelligence and information on a permanent and continuous basis, every single day of the year.

In this context, NATO may take the lead or play a supporting role in the context of a crisis management activity undertaken under the responsibility of the United Nations, the OSCE, the European Union, or by one or more NATO member countries. As the Alliance says, in either case, the focus of NATO's involvement is on making a significant and distinct contribution to successful conflict management and resolution.

**The crisis management process**

In this process the Alliance must be prepared to conduct the full range of Article 5 and non-Article 5 missions in circumstances that in many cases will be difficult to predict since, to some extent, every crisis is unique. Nonetheless, the process by which the Alliance addresses and seeks to manage and resolve a crisis can be planned with reasonable confidence. The crisis management process is designed to facilitate political consultation and decision-making at a sufficiently early stage in an emerging crisis to give the appropriate NATO committees time to coordinate their work and submit timely advice to the Council. It also allows the Supreme Allied Commander Europe, as the Strategic Commander responsible for Allied Command Operations, to undertake preparatory planning measures in a reasonable timeframe. These activities may in turn contribute early on to the advice provided to the Council by NATO's military authorities.

Thus, in an emerging crisis calling for possible crisis response operations, the crisis management process consists of five successive phases ranging from initial indications and warning of an impending crisis, assessment of the situation and its actual or potential implications for Alliance security, development of recommended response options, and planning and execution of the Council's decisions.

In this way, the effectiveness of the crisis response system and of NATO's overall crisis management process may be determined to a great extent by the effectiveness and efficiency of the structures and procedures of the NATO Headquarters Crisis Management Organization, which have to be responsive, flexible and adaptable. They must also facilitate the seamless and smooth inter-operation of the other main elements of the crisis management process, namely the NATO Crisis Response System (NCRS), the NATO Intelligence and Warning System (NIWS), NATO's operational planning system, and NATO Civil Emergency Planning crisis management arrangements. The NATO

Situation Centre supports the process with communications and other essential facilities.

It should be noted that in the light of decisions taken at the Washington Summit meeting in 1999 to transform NATO structures and capabilities, the crisis management tools in place were considered to be no longer sufficiently well adapted to the risks and challenges that the Alliance might face. Accordingly, in August 2001, the North Atlantic Council approved policy guidelines with a view to developing a single, fully integrated NATO Crisis Response System (NCRS).

The terrorist attacks on the United States in September 2001 brought new urgency to this task and a new dimension to the NATO's crisis management framework, which had hitherto focused primarily on requirements for collective defence. In June 2002, the Council also provided political guidance for the development of a Military Concept for Defence Against Terrorism. An important result of this decision has been the introduction of measures to strengthen civil emergency planning for Article 5 and non-Article 5 contingencies, as well as the management of the consequences of civil emergencies or disasters resulting from the use of chemical, biological, radiological or nuclear (CBRN) agents. That is to say, this Concept foresaw a number of new initiatives, such as intelligence sharing, CBRN measures, the establishment of a Terrorist Threat Intelligence Unit, and Civil Emergency planning, as a priority. Yet all these separate initiatives lacked coordination and an overarching vision.

In view of new risks, as well as the need for the Alliance to be able to address more complex and demanding crisis management requirements, including the possibility of

NATO support for non-NATO operations involving one or more member countries, further far-reaching decisions have been taken with regard to NATO's overall defence posture. These have resulted in a new force posture and a new command structure, transformation of staff structures, new measures relating to defence against terrorism, the establishment of the NATO Response Force, improvements in capabilities, the development of the strategic partnership with the European Union, enhanced cooperation with Partner countries, and reinforcement of the Alliance's Mediterranean Dialogue.

Even more, the NATO Crisis Response System under development takes full account of, and complements these new NATO concepts, capabilities and arrangements. It aims to provide the Alliance with a comprehensive set of options and measures to manage and respond to crises appropriately, taking full advantage of the tools and capabilities being introduced as a result of decisions taken by NATO heads of state and government at successive summit meetings.

We can complete the above by saying that exercises to test and develop crisis management procedures are held at regular intervals in conjunction with national capitals and NATO Strategic Commanders. Such exercises and the arrangements, procedures and facilities on which the crisis management process depends are coordinated by the Council Operations and Exercise Committee (COEC).

Besides, crisis management activities involving NATO's Partner countries are also coordinated by the COEC and are among the agreed fields of activity in the Euro-Atlantic Partnership Work Plan and in Individual Partnership Programmes. They include briefings and consultations, expert visits, crisis management courses, Partner country participation in an annual NATO-wide crisis management exercise, and the provision of generic crisis management documents to interested Partner countries.

Finally, the coordination of crisis management responses to disasters or emergencies in the Euro-Atlantic area takes place in the framework of the Euro-Atlantic Disaster Response Coordination Centre (EADRCC). The Centre's role is to facilitate the coordination of responses to civil emergencies or disasters, including the management of consequences resulting from terrorist attacks. The Centre, which can be augmented if necessary, is able to operate on a 24/7 basis if circumstances require.

# THE DEFENCE PLANNING DIMENSION

In the current political and strategic environment in Europe, the success of the NATO's role in preserving peace and preventing war depends, even more than in the past, on the effectiveness of preventive diplomacy and the successful management of crises that affect security. The political, economic, social and environmental elements of security and stability are becoming increasingly important.

However, the defence dimension of the Alliance is the concrete expression of the Alliance's general deterrent role in defending its member countries and thus contributes to the maintenance of stability in Europe. The maintenance of an adequate military capability and clear preparedness to act collectively in the common defence remain central to the Alliance's security objectives. Ultimately this capability, combined with political solidarity, is designed to prevent any attempt at coercion or intimidation, and to ensure that military aggression directed at the Alliance can never be perceived as an option with any prospect of success, thus guaranteeing the security and territorial integrity of member states and protecting Europe as a whole from the consequences which would ensue from any threat to the Alliance.

At the same time, it is imperative to note that defence planning is the basis for all NATO's crisis management and military operations. Its scope has been extended to enable NATO to react to a much wider range of contingencies than in the past and Alliance forces have been radically reorganised in order to enable the full range of defence policy and plans, from conventional deterrence to conflict resolution, peace support, humanitarian intervention and other operational tasks to be fulfilled.

Fundamental is that the framework for NATO's defence planning process is provided by the underlying principles which are the basis for collective security as a whole: political solidarity among member countries, the promotion of collaboration and strong ties between them in all fields where this serves their common and individual interests, the sharing of roles and responsibilities and recognition of mutual commitments, and a joint undertaking to maintain adequate military forces to support Alliance strategy and policy.

To determine the size and nature of their contribution to collective defence, NATO member countries retain full sovereignty and independence of action. However, the nature of NATO's defence structure requires that member countries take into account the overall needs of the Alliance when making their individual decisions.

Therefore they follow agreed defence planning procedures which provide the methodology and machinery for determining the forces needed for the implementation of Alliance policies, for coordinating national defence plans and for establishing force planning goals which are in the interests of the Alliance as a whole.

Thus, the planning process takes many factors into account, including changing political circumstances, assessments provided by NATO's strategic military commanders of the forces required to fulfil their tasks, technological developments, the importance of an equitable division of roles, risks and responsibilities within the Alliance, and the individual economic and financial capabilities of member countries. The process thus ensures that all relevant considerations are jointly examined to enable the best use to be made of collective national resources which are available for NATO roles.

In this sense, close coordination between international civil and military staffs, NATO's military authorities, and governments is maintained through an annual exchange of information on national plans. This exchange of information enables each country's intentions to be compared with NATO's overall requirements and, if necessary, to be reconsidered in the light of new ministerial political directives, modernisation requirements and changes in the roles and responsibilities of the forces themselves. All these aspects are kept under continual review and are scrutinised at each stage of the defence planning cycle.

## Defence planning process

According to a 2004 Istanbul Summit some changes were done in the Alliance's planning processes in order to answer to current and future operational requirements.

The agreed changes support the further transformation of Alliance military capabilities through a coherent and streamlined process designed to ensure that NATO continues to develop the forces and capabilities needed to conduct the full range of Alliance missions. This includes providing support for operations which might be led by the European Union in the context of the European Security and Defence Identity and its strategic partnership with NATO. Also in that context, the process enables all European Allies to benefit from NATO support in the context of their operational planning for the conduct of EU-led operations.

As we have seen previously, since 1991 the starting point for defence planning has been the Alliance's Strategic Concept, which sets out in broad terms Alliance objectives and the means for achieving them. The original Strategic Concept has been superseded by the Alliance's new Strategic Concept approved by NATO heads of state at their Washington Summit meeting in April 1999. The review of defence planning conducted by the Defence Review Committee during 2003 and 2004, which resulted in changes designed to facilitate the transformation of NATO's military capabilities, also takes the Strategic Concept as its starting point, together with the development of the Alliance's new tasks and challenges and the evolution of the security environment as a whole. Finally, the Strategic Concept took its final form in 2010, as we have previously explained.

According to NATO philosophy, while defence planning in the broadest sense embraces a wide spectrum of planning disciplines ranging from force and armaments planning to aspects such as logistics, standardisation, nuclear planning, communications, civil emergency planning, air defence, and resource planning, the area of defence planning examined in the course of the above review encompasses NATO's force planning procedures and their relationship with these disciplines. The role of defence planning in this context is to provide a framework which permits national and multinational defence planning arrangements to be harmonised in order to meet the Alliance's agreed requirements in the most effective way. The aim is to ensure the availability of national forces and capabilities required for the full range of Alliance missions by setting targets for implementation and assessing the degree to which these targets are being met.

And once again following the provisions of the guidelines of the Alliance, the conclusions of the review recommended the retention of the basic principles of the defence planning process as it has evolved, including its threepillar structure. This is based, firstly, on overall political guidance, secondly on the adoption of agreed planning targets to fulfil the objectives established in the guidance, and lastly on a systematic review process to monitor, and where necessary adjust or correct, the implementation of the targets.

However, changes have been introduced that affect the duration of the planning cycle and the periodicity of the steps it involves. Changes in the procedures for the development of political guidance and the levels at which it is drawn up have also been made, introducing a distinction between the Comprehensive Political Guidance agreed upon at a high level and the more detailed guidance routinely elaborated as part of the normal procedures of the defence planning process within NATO under the authority of the North Atlantic Council or the Defence Planning Committee.

Political guidance will include consideration of a concept known as NATO's "level of ambition". This refers to the agreed assessment by the member governments of the number, scale and nature of operations that NATO should be able to conduct. With regard to force planning, in addition to that assessment, political guidance also encompasses the guidance agreed by defence ministers meeting in the Defence Planning Committee and supplementary guidance that may be agreed by the Defence

Planning Committee meeting in permanent session at the level of ambassadors.

Two further specific areas covered by the review and leading to changes in the defence planning process should also be mentioned.

1) Firstly, the review allows for the incorporation, within the planning procedures, of measures to enhance cooperation between NATO and the European Union in the field of defence planning and the improvement of capabilities. This is designed to enable the question of the availability of forces for EU-led operations to be addressed in a more comprehensive manner.

2) Secondly, the review recognises the need for better coordination and harmonisation of all defence planning disciplines across the board and includes provision for further work to be done in appropriate areas to bring this about.

In accordance with the review's recommendations, the guidance required as the first step in the process is issued by defence ministers every four years, with the possibility of a biennial update if necessary, in a document known as "Ministerial Guidance". This gives guidance on defence planning in general and force planning in particular, reflecting political, economic, technological and military factors which could affect the development of forces and capabilities, and their strategic implications. It sets out the priorities and areas of concern to be addressed by the NATO Military Authorities in drawing up their force goals in the first instance, and secondly by countries in their own planning. It deals with planning for forces and capabilities required both for collective defence and for contingencies falling outside the scope of Article 5 of the Washington Treaty. It may also provide guidance on cooperation with other organisations and, since 1997, has included political guidance defining the likely scope of European-led operations.


**Planning targets and force goals**

It is basic to know that, specific planning targets for the armed forces of each member country are developed on the basis of ministerial guidance. These targets, for which the starting point is the identification of military requirements by the NATO Strategic Commands, incorporate NATO force goals developed from draft force proposals put forward by Allied Command Transformation and designed to enable Allied Command Operations to accomplish the full range of operational missions that may be assigned to it by the North Atlantic Council. The draft proposals are subsequently discussed with individual nations and if necessary amended, prior to being examined collectively by the NATO Military Committee.

That examination takes into account the military validity and technical feasibility of the proposals and, based on the Military Committee's conclusions, results in draft force goals that are then submitted to the Defence Planning Committee for its approval and formal adoption as NATO force goals. The force goals may be complemented in some cases by reinvestment goals, which are drawn up in response to requests by member governments. These combine, on the one hand, the identification of force elements that are no longer needed to meet Alliance requirements and can be eliminated and, on the other hand, the identification of other priority capabilities which may be met by the resources thus freed.

Let us point out that, the goals generally cover a four-year period but in certain cases look further into the future. The procedures also make provision for the goals to be updated when circumstances require, normally at the mid-point of the planning cycle.

## The defence review

In the tangle of concepts and structures of the Alliance, let us emphasise that the third leg of the force planning cycle is the defence review process that takes place every second year and is conducted during a period of a little over twelve months. It consists of the individual and collective scrutiny and assessment of the force plans and corresponding financial planning of individual member countries, measured against the yardstick of the agreed NATO force goals for a ten-year planning period.

The defence review serves two purposes. It allows an assessment to be made of the degree to which individual countries are meeting their targets in terms of NATO's force goals, output targets and national usability targets. It also enables an assessment of the extent to which combined Alliance military forces and capabilities are able to meet the political guidelines issued for the current planning cycle. These assessments represent both a measurement and a corrective mechanism, allowing shortcomings to be highlighted as well as areas where increased multinational cooperation may offer advantages. More generally, the assessments provide an evaluation of the extent to which the burden of contributing to Alliance capabilities and military operations is equitably distributed among the member countries.

The conduct of the defence review itself draws on well-established mechanisms beginning with the issuing of a Defence Planning Questionnaire and the analysis of responses to it, resulting in draft Country Chapters based on inputs from NATO's international defence planning staff and from the two Strategic Commands. Following a trilateral meeting with each member country, normally taking place in the respective capitals, revised Country Chapters are subjected to a multilateral examination at the level of the Defence Review Committee. This aims in particular at reconciling possible differences between national and NATO force goals or plans. When this examination has been completed, the Country Chapters are transformed into individual national annexes to a general report to be submitted to the Defence Planning Committee at its spring ministerial meeting.

Thus, this process is repeated for each member country participating in NATO's integrated military structure, over a period of several months, culminating in the preparation of a General Report. The latter also includes a report by the Military Committee on the military suitability of the emerging NATO Force Plan and on the degree of military risk associated with it. Finally, the General Report contains a section coordinated with relevant bodies of the European Union, and based on the contributions of relevant European member countries, setting out the extent to which the emerging plan can be expected to meet EU force and capability requirements.

It is important to highlight that, the overall force planning process may contain one further element in the form of an Overall Summary Appraisal of Defence Planning which may be presented at any time by the NATO Secretary General, giving his view of the current and future state of Alliance defence and of its force plans.The appraisal may serve to highlight points relating to specific national plans, identify issues that may need to be discussed by defence ministers, and help to establish links between different spheres of defence planning that might not otherwise be considered in relation to each other.

Many of the above elements of NATO's defence and force planning procedures are increasingly being used within the Partnership for Peace structure as a means of enhancing interoperability between the military structures of NATO and its Partner countries, assisting the process of defence reform within Partner countries and facilitating the participation of Partner countries in NATO-led operations.

# THE ECONOMIC DIMENSION

Let us start by saying that the basis for economic cooperation within the Alliance is Article 2 of the North Atlantic Treaty, which states that member countries "will seek to eliminate conflict in their international economic policies and will encourage economic collaboration between any or all of them".

We must keep in mind that NATO's core business is security and defence, so its work in the economic field is focused on specific economic issues relating to security and defence where it can offer added value.

Accordingly, the Organization reinforces collaboration between its members whenever economic issues of special interest to the Alliance are involved. This applies particularly to issues which have direct security and defence implications. The Alliance acts as a forum in which different and interrelated aspects of political, military and economic questions can be examined.

The NATO Economic Committee is the only Alliance forum concerned exclusively with consultations on economic developments with a direct bearing on security policy. It meets in different formations and is supported by the Defence and Security Economics Directorate of the Political Affairs and Security Policy Division of NATO's International Staff.

In the context of the Alliance's overall security interests and in line with its evolving priorities, the work of the Committee covers a wide range of issues and regularly involves the preparation of analyses and assessments relating to NATO's political and security agenda.

Close cooperation is maintained with a network of experts from capitals, enabling the Directorate to serve as a unique forum for sharing information and expertise on defence and security economic issues related to countries and regions of concern to NATO and to areas where NATO is playing an operational role. The economic and financial dimensions of terrorism have become a firm part of this agenda. Based on contributions provided by member countries, agreed assessments of economic intelligence matters are regularly produced for the benefit of the North Atlantic Council, Allied capitals and military bodies.

Besides, the Defence and Security Economics Directorate is also involved in monitoring both general economic and defence economic aspects of the Membership Action Plan such as the affordability and sustainability of defence spending.

We want to bring up that another significant facet of NATO's economic dimension is its cooperative activities with Partner and other countries with which the Alliance has developed cooperative relations, including security and defence economic work carried out in the framework of the Euro-Atlantic Partnership Council, the NATO-Russia Council, the NATO-Ukraine Commission, and NATO's relations with South East Europe Initiative, Mediterranean Dialogue and Istanbul Cooperation Initiative countries. This includes economic aspects of defence budgeting and resource management in defence spending, defence conversion matters (for example relating to retraining of military personnel and conversion of military sites and defence industries), economic aspects of the international fight against terrorism and other relevant economic security issues.

We should also quote here cooperation in the context of the Euro-Atlantic Partnership Council takes place through conferences, workshops and experts meetings. Joint cooperation schemes have also been developed in association with external institutions such as the George C. Marshall Center for Security Studies. These mechanisms have enabled the experience of NATO countries to be made available to Partner countries in a number of fields, recent examples of which have included economic dimensions of defence institution-building, economic and financial aspects of terrorism, economic aspects of security and defence in the Southern Caucasus, and new techniques for managing defence resources in Allied and Partner countries. The Directorate also monitors defence and security economic issues included in Individual Partnership Action Plans.

Besides, cooperation with Russia in the framework of the NATO-Russia Council's Ad Hoc Working Group on Defence Reform is focused in the first instance on expert-level exchanges on a wide spectrum of topics ranging from macroeconomic, financial, budgetary and social aspects of defence reform to the restructuring of defence industries. Secondly, a Memorandum of Agreement was signed with the Russian Ministry of Defence in June 2001 on the opening of a NATO-Russia Information, Consultation and Training Centre for the resettlement of military personnel due for discharge or discharged from the Russian Federation armed forces. This Centre, which operates in the six Russian military districts, is financially supported by NATO and organises training courses, "train the trainer" courses and meetings of experts on current topics. The Centre also runs a comprehensive website including a wide range of practical information for released military personnel. The Centre's work is a very concrete and practical example of cooperation between NATO and the Russian Ministry of Defence.

We could also say that specific activities in the area of economic cooperation are also conducted within the framework of the NATO-Ukraine Annual Target Plans. They include meetings of the Joint Working Group on Economic Security, courses on economic aspects of the defence budgetary process, exchanges on the restructuring of defence industries and social issues relating to defence reform. There are also regular consultations on general economic policy and on structural and macro-economic trends in Ukraine. Since 1999, NATO has financed retraining courses in various cities of Ukraine, which have facilitated training in foreign languages and in management techniques for some one hundred former Ukrainian military officers each year. This programme has produced tangible benefits, greatly facilitating the reintegration process for released military personnel.

Finally, comprehensive programmes on the retraining of released military personnel and military base conversion in south-eastern Europe are also monitored by the Defence and Security Economics Directorate. NATO has taken the lead on these issues in the framework of the Stability Pact for South Eastern Europe. Through teams of experts from Allied and Partner countries led by the Directorate, NATO has provided advice to a series of countries for the development of appropriate reconversion programmes adapted to their needs. The teams make available expertise, technical assistance and recommendations, based on general experience and taking into account the specific situation facing the countries concerned. NATO's work in this field contributes substantively to the difficult process of defence reform and conversion in the region. Defence conversion schemes worked out with NATO's assistance have demonstrated their worth as blueprints for project implementation.

# COMMON-FUNDED RESOURCES, BUDGETS AND FINANCIAL MANAGEMENT

We will start by saying that NATO is an intergovernmental organisation to which member countries allocate the resources needed to enable it to function on a day-to-day basis and to provide the facilities required for consultation, decision-making and the subsequent implementation of agreed policies and activities. It is supported by a military structure which provides for the common defence of the member countries, cooperation with NATO's Partner countries and implementation of Alliance policies in peacekeeping and other fields. Since NATO has only a limited number of permanent headquarters and small standing forces, the greater part of each member country's contribution to NATO, in terms of resources, comes indirectly through its expenditure on its own national armed forces and on its efforts to make them interoperable with those of other members so that they can participate in multinational operations. Member countries also incur the deployment costs involved whenever they volunteer forces to participate in NATO-led operations.

Hence, with few exceptions, NATO funding does not cover the procurement of military forces or of physical military assets such as ships, submarines, aircraft, tanks, artillery or weapon systems. An important exception is the NATO Airborne Early Warning and Control Force, a fleet of radar-bearing aircraft jointly procured, owned, maintained and operated by member countries and placed under the operational command and control of a NATO Force Commander responsible to the NATO Strategic Commanders. NATO also finances investments directed towards collective requirements, such as air defence, command and control systems or Alliance-wide communications systems which cannot be designated as being within the responsibility of any single member country to provide. Such investments are subject to maintenance, renewal and ultimately replacement in accordance with changing requirements and technological developments. The expenditures this requires also represent a significant portion of NATO funding.

Thus, member countries make direct contributions to budgets managed directly by NATO, in accordance with an agreed cost-sharing formula broadly calculated in relation to their ability to pay. These contributions represent a small percentage of each member's overall defence budget and, as a general rule, finance the expenditures of those parts of the NATO structure in which they participate. These contributions, made within the framework of NATO, often follow the principle of common funding.

Also projects can be jointly funded, which means that the participating countries can identify the requirements, the priorities and the funding arrangements, but NATO has visibility and provides political and financial oversight. Joint funding arrangements typically lead to the setting-up of a management organisation and an implementation agency in areas such as aircraft and helicopter production, air defence and logistics. Additionally, NATO member countries can cooperate within the framework of NATO on an ad hoc basis for a range of other, more limited, activities. This cooperation can take the form of trust fund arrangements, contributions in kind, ad hoc cost sharing arrangements and donations.

## Common funding

As mentioned earlier, the large majority of resources are national. NATO resource planning aims to provide the Alliance with the capabilities it needs, but focuses on the elements that are joined in common funding, that is to say where member pool resources within a NATO framework. When a need for expenditure has been identified, discussions take place among the potential contributing countries to determine whether the principle of common funding should be applied. In other words whether the requirement serves the interests of all the contributing countries and therefore should be borne collectively.

The common funding structure is diverse and decentralised. Certain multinational cooperative activities relating to research, development, production and logistic support do not involve all and, in some instances, may only involve a small number of member countries. These activities, most of which are managed by NATO Production and Logistics Organizations, are subject to the general financial and audit regulations of NATO but otherwise operate in virtual autonomy under charters granted by the North Atlantic Council.

Thus, the criteria for common funding are held under constant review and changes may be introduced as a result of new contingencies. For example, the need to develop clear definitions of the parts of NATO's crisis response costs which should be imputed to international budgets and those which should be financed by national budgets. Other changes may result from organisational or technological developments or simply from the need to control costs in order to meet requirements within specific funding limitations.

Despite those challenges, the principle of common funding on the basis of consensus remains fundamental to the workings of the Alliance.

Common funding arrangements principally include the NATO Civil and Military Budgets, as well as the NATO Security Investment Programme (NSIP). These are the only funds where NATO authorities identify the requirements and set the priorities in line with overarching Alliance objectives and priorities.

1.- The Civil Budget:

The Civil Budget is formulated on an objective-based framework, which establishes clear links between NATO's Strategic Objectives and the resources required to achieve them. It provides funds for personnel expenses, operating costs, and capital and programme expenditure required to achieve four frontline objectives and three support objectives.

The frontline objectives are:

- providing effective policy, planning and resourcing in support of NATO operations in the Euro-Atlantic region and beyond;

- conducting necessary policy and planning work to promote and support improved Alliance capabilities;

- supporting consultation and cooperative activities with partners to strengthen security and respond to new security challenges and threats to the Euro-Atlantic region;

- building awareness of, and support for, NATO, its operations, and its role in promoting security through public diplomacy.

2.- The Military Budget:

The Military Budget covers the operating and maintenance costs of the international military structure. This includes, for instance, the Military Committee, the International Military Staff, military agencies, the two strategic commands and associated command, control and information systems, research and development agencies and the NATO Airborne Early Warning and Control Force. The military budget also covers the operating costs of the command structures for crisis response operations and missions undertaken by NATO.

It is funded primarily by the ministries of defence of each member country, supervised by the Military Budget Committee and implemented by the individual budget holders.

3.- The NATO Security Investment Programme:

Also NATO member countries contribute to the NATO Security Investment Programme (NSIP). This covers major construction and command and control system investments needed to support the roles of the NATO strategic commands, but which are beyond the national defence requirements of individual member countries. Both the

Military Budget and the NSIP, are guided by the "over and above" rule: "Common funding will focus on the provision of requirements which are over and above those which could reasonably be expected to be made available from national resources". The NSIP includes, for example, requirements for crisis response operations and military installations and capabilities such as communications and information systems, air command and control systems, satellite communications, military headquarters, airfields, fuel pipelines and storage, harbours, and navigational aids.

The NSIP is financed by the ministries of defence of each member country and is supervised by the Infrastructure Committee. Projects are implemented either by individual host countries or by different NATO agencies and strategic commands, according to their area of expertise.

4.- Resource management:

Under pressures to optimise the allocation of military common-funded resources, since the mid1990s, member countries have reinforced NATO's management structure by promoting the development of capability packages and by establishing the Senior Resource Board (SRB) which has responsibility for overall resource management of NATO's military resources (for example, excluding resources covered by the Civil Budget).

The capability packages identify the assets available to and required by NATO military commanders to fulfil specified tasks. They are a prime means of assessing common-funded supplements (in terms of both capital investment and recurrent operating and maintenance costs) as well as the civilian and military manpower required to accomplish the task. These packages are reviewed by the Senior Resource Board composed of national representatives, representatives of the Military Committee and the NATO Strategic Commanders and the Chairmen of the Military Budget, Infrastructure and NATO Defence Manpower Committees.

The Board endorses the capability packages from the point of view of their resource implications prior to their approval by the North Atlantic Council or Defence Planning Committee as applicable. It also annually recommends for approval by the North Atlantic Council a comprehensive Medium Term Resource Plan which sets financial ceilings for the following year and planning figures for the four subsequent years.

Within these parameters the Military Budget and Infrastructure and Defence Manpower Committees oversee the preparation and execution of their respective budgets and plans. The Board further produces an Annual Report which allows the North Atlantic Council to monitor the adequacy of resource allocations in relation to requirements and to review the military common-funded resource implications for NATO's common-funded budgets of new Alliance policies.


**Financial management**

In this section we have to review that financial management within NATO is structured to ensure that the ultimate control of expenditure rests with the member countries supporting the cost of a defined activity, and is subject to consensus among them.

This control is exercised, at all levels of decision-making, either in terms of general limitations (for example, allocation of fixed resources for operating costs), or by specific restrictions (for example, temporary immobilisation of credits or the imposition of specific economy measure). These controls may be stipulated in the terms in which approval of the budget is given or exercised by contributing countries through exceptional interventions in the course of the execution of the budget. The financial managers, such as the Secretary General, NATO Strategic Commanders and Subordinate Commanders and other designated Heads of NATO bodies, have relative discretion to propose and execute their budgets.

Besides, no single body exercises direct managerial control over all four of the principal elements of the Organization's financial structure: 1) the International Staff (financed by the Civil Budget); 2) the international military structure (financed by the Military Budget); 3) the Security Investment Programme; and 4) specialised Production and Logistics Organizations.

The latter fall into two groups: a) those which are financed under arrangements applying to the international military structure; and b) those which operate under charters granted by the North Atlantic Council, with their own Boards of Directors and finance committees and distinct sources of financing within national treasuries.

Financial management of the organisational budgets:

In NATO the financial management of the Civil and Military Budgets differ from that of the Security Investment Programme. Financial regulations provide basic unifying principles around which the overall financial structure is articulated. They are approved by the North Atlantic Council, and are complemented by rules and procedures adapting them to specific NATO bodies and programmes.

Here the budget is annual, coinciding with the calendar year. It is prepared under the authority of the Head of the respective NATO body, reviewed and recommended for approval on the basis of consensus by a finance committee composed of representatives of contributing member countries, and approved for execution by the North Atlantic Council. Failure to achieve consensus before the start of the financial year entails non-approval of the budget and the financing of operations, under the supervision of the finance committee, through provisional allocations limited to the level of the budget approved for the preceding year. This regime may last for six months, after which the Council is required to decide either to approve the budget or to authorise continuation of interim financing.

Thus, then the budget has been approved, the Head of the NATO body has discretion to execute it through the commitment and expenditure of funds for the purposes authorised. This discretion is limited by different levels of constraint prescribed by the Financial Regulations regarding such matters as recourse to competitive bidding for contracts for the supply of goods and services, or transfers of credits to correct over or under-estimates of the funding required.

Discretionary authority to execute a budget may be further limited by particular obligations to seek prior approval for commitments and expenditure. These may occasionally be imposed by the finance committee in the interests of ensuring strict application of new policies or of monitoring the implementation of complex initiatives such as organisational restructuring.

Financial management of the NATO Security Investment Programme:

In this matter implementation of the NATO Security Investment Programme has its starting point in the capability packages. Once these have been approved, authorisation of individual projects can commence under the responsibility of the Infrastructure Committee. The Host Nation (either the country on whose territory the project is to be implemented, a NATO agency or a strategic command) prepares an authorisation request. Once the Committee has agreed to the project, the Host Nation can proceed with its final design, contract award and implementation. Unless otherwise agreed by the Infrastructure Committee, the bidding process is conducted among firms from those countries contributing to the project.

The financial management system which applies to the NSIP is based on an international financial clearing process. Host nations report on the expenditure foreseen on authorised projects within their responsibility. Following agreement of the forecasts by the Infrastructure Committee, the International Staff calculates the amounts to be paid by each country and to be received by each host nation. Further calculations determine the payment amounts, currencies and which nation or NATO agency will receive the funds, these are computed on a quarterly basis. Once a project has been completed, it is subject to a Joint Final Acceptance Inspection to ensure that the work undertaken is in accordance with the scope of work authorised. As soon as this report is accepted by the Infrastructure Committee, it is added to the NATO inventory.

So there are several levels of financial reporting. Twice a year the International Staff prepares for each Host Nation Semi-Annual Financial Reports on projects under implementation. Quarterly, the pre-paysheet and paysheet are published. These reports refer to the transfer of funds between host nations. An NSIP Expenditure Profile is prepared every spring, which covers the NSIP expenditure levels for the next 10 years. The NSIP Financial Statements are prepared in the spring of each year. They portray the financial situation of the NSIP as at 31 December of each year and the summary of activity during the year. These statements serve as the baseline for Infrastructure Committee discussion on the state of the NSIP.


**Financial control**

Here with respect to the Military Budget and the Civil Budget, the head of the respective NATO body is ultimately responsible for the correct preparation and execution of the budget, the administrative support for this task is largely entrusted to his Financial Controller.

The appointment of this official is the prerogative of the North Atlantic Council, although the latter may delegate this task to the relevant finance committee. Each Financial Controller has final recourse to the finance committee in the case of persistent disagreement with the Head of the respective NATO body regarding an intended transaction. The responsibility for the management of the NSIP finances rests with the Controller for Infrastructure. Through a professional staff, he exercises financial control and implementation oversight.

It is important to know that the Financial Controller is charged with ensuring that all aspects of execution of the budget conform to expenditure authorisations, to any special controls imposed by the finance committee and to the Financial Regulations and their associated implementing rules and procedures. He may also, in response to internal auditing, install such additional controls and procedures as he deems necessary for maintaining accountability.

A major task of the NATO strategic commands' Financial Controllers is to ensure that the funds required to finance execution of the budget are periodically called up from contributing member countries in accordance with their agreed cost shares and in amounts calculated to avoid the accumulation of excessive cash holdings in the international treasury. The outcome of all these activities is reflected in annual financial statements prepared and presented for verification to the International Board of Auditors.

An important point is that an independent International Board of Auditors for NATO is responsible for auditing the accounts of the different NATO bodies and the efficiency and effectiveness of their operations from a financial perspective as well as for auditing expenditure under the NATO Security Investment Programme.

The Board's mandate includes not only financial but performance audits, therefore extending its role beyond safeguarding accountability to the review of management practices in general. The Board is composed of officials normally drawn from the national audit bodies of member countries appointed by Council and responsible for their work only to the Council. The principal task of the Board is to provide the North Atlantic Council and member governments with the assurance that common funds are properly used for the settlement of authorised expenditure and that expenditure is within the physical and financial authorisations granted.

**NUCLEAR POLICY**

In this matter it is essential to know that NATO's nuclear strategy and force posture are inseparable elements of the Alliance's overall strategy of war prevention. They fulfil a fundamentally political role in preserving peace and contributing to stability in the Euro-Atlantic region. However, under the momentous security improvements which have been achieved since the end of the Cold War, the Alliance has been able to reduce its reliance on nuclear forces radically. NATO's nuclear powers (France, the United Kingdom and the United States) took unilateral steps to cancel planned modernisation programmes for their nuclear forces.

Moreover, the Alliance's strategy, while remaining one of war prevention, is no longer dominated by the possibility of escalation involving nuclear weapons and its nuclear forces no longer target any country. Among the steps taken to adapt to the new security environment, the changes to the nuclear elements of its strategy and force posture were among the first and most incisive measures

All the more, NATO's nuclear forces contribute to European peace and stability by underscoring the irrationality of a major war in the Euro-Atlantic region. They make the risks of aggression against NATO incalculable and unacceptable in a way that conventional forces alone cannot. They also create uncertainty for any country that might contemplate seeking political or military advantage through the threat or use of nuclear, biological or chemical (NBC) weapons against the Alliance. By promoting European stability, helping to discourage threats relating to the use of weapons of mass destruction, and contributing to deterrence against such use, NATO's nuclear posture serves the interests not only of the NATO Allies but also of its Partner countries and of Europe as a whole.

NATO's reduced reliance on nuclear forces has been manifested in major reductions in the forces themselves. In 1991 NATO decided to reduce the number of weapons which had been maintained for its sub-strategic forces in Europe by over 85 per cent compared to Cold War levels. In addition to the reductions of sub-strategic forces, the strategic forces available to the NATO Allies have also been dramatically reduced.

It is important to know that the terms "strategic" and "sub-strategic" have slightly different meanings in different countries. Strategic nuclear weapons are normally defined as weapons of "intercontinental" range (over 5500 kilometres), but in some contexts these may also include intermediate-range ballistic missiles of lower ranges. The term "sub-strategic" nuclear weapons has been used in NATO documents since 1989 with reference to intermediate and short-range nuclear weapons and now refers primarily to air-delivered weapons for NATO's dual-capable aircraft and to a small number of United Kingdom Trident warheads in a sub-strategic role (other sub-strategic nuclear weapons having been withdrawn from Europe and subsequently eliminated).

In this sense, the only land-based nuclear weapons which NATO retains in Europe are gravity bombs for dual-capable aircraft. These weapons have also been substantially reduced in number and are stored in a smaller number of locations in highly secure conditions. The readiness levels of dual-capable aircraft associated with them have been progressively reduced, and increased emphasis has been placed on their conventional roles. None of these nuclear weapons are targeted against any country.

Continuing with the doctrine maintained by NATO in this matter, the NATO Allies have judged that the Alliance's requirements can be met, for the foreseeable future, by this "sub-strategic" force posture. NATO has also declared that enlarging the Alliance will not require a change in its current nuclear posture. NATO countries have no intention, no plan and no reason to deploy nuclear weapons on the territory of new members, nor any need to change any aspect of NATO's nuclear posture or nuclear policy, and they do not foresee any future need to do so.

Thus, the collective security provided by NATO's nuclear posture is shared among all members of the Alliance, providing reassurance to any member that might otherwise feel vulnerable. The presence of US nuclear forces based in Europe and committed to NATO provides an essential political and military link between the European and North American members of the Alliance. At the same time, the participation of non-nuclear countries in the Alliance nuclear posture demonstrates Alliance solidarity, the common commitment of its member countries to maintaining their security and the widespread sharing among them of burdens and risks

Finally, we underline that political oversight of policies dictating NATO's nuclear posture is also shared among member countries. NATO's Nuclear Planning Group provides a forum in which the defence ministers of nuclear and non-nuclear Allies take part in the development of the

Alliance's nuclear policy and in decisions on NATO's nuclear posture.

## NATO and the INF Treaty

The Intermediate-Range Nuclear Forces Treaty, or INF Treaty, was crucial to Euro-Atlantic security for decades. It eliminated a whole category of nuclear weapons that threatened Europe in the 1980s. All NATO Allies agree that the SSC-8/9M729 missile system developed and deployed by Russia violated the INF Treaty, while posing a significant risk to Alliance security. Despite Allies' repeated calls on Russia to return to full and verifiable compliance, Russia continued to develop and deploy Treaty-violating systems, which led to the agreement's demise on 2 August 2019.

The INF Treaty was signed on 8 December 1987 by the United States and the former Soviet Union, and entered into force on 1 June 1988. It required both countries to eliminate their ground-launched ballistic and cruise missiles that could travel between 500 and 5,500 kilometres (between 300 and 3,400 miles) by an implementation deadline of 1 June 1991.

By the deadline, the two countries had together destroyed a total of 2,692 short- and intermediate-range missiles: 1,846 Soviet missiles and 846 American missiles. It marked the first elimination of an entire category of weapons capable of carrying nuclear warheads.

But in recent years, Russia has developed, produced, tested and deployed a new intermediate-range missile known as the 9M729, or SSC-8. The 9M729 is mobile and easy to hide. It is capable of carrying nuclear warheads. It reduces warning times to minutes, lowering the threshold for nuclear conflict. And it can reach European capitals.

In July 2018, NATO Allies stated that after years of denials and obfuscation by the Russian Federation, and despite Allies repeatedly raising their concerns, the Russian Federation had only recently acknowledged the existence of the missile system without providing the necessary transparency or explanation. A pattern of behaviour and information over many years led to widespread doubts about Russian compliance. NATO Allies said that, in the absence of any credible answer from Russia on this new missile, the most plausible assessment was that Russia was in violation of the Treaty.

In December 2018, NATO Foreign Ministers supported the finding of the United States that Russia was in material breach of its obligations under the INF Treaty and called on Russia to urgently return to full and verifiable compliance with the Treaty.

Allies remained open to dialogue and engaged Russia on its violation, including at a NATO-Russia Council meeting on 25 January 2019. Russia continued to deny its INF Treaty violation, refused to provide any credible response, and took no demonstrable steps toward returning to full and verifiable compliance.

As a result of Russia's continued non-compliance, on 1 February 2019, the United States announced its decision to suspend its obligations under Article XV of the INF Treaty. This meant that the United States could terminate the Treaty within six months of this date if Russia had not come back into compliance.

Also on 1 February 2019, NATO Allies said that unless Russia honoured its INF Treaty obligations through the verifiable destruction of all of its 9M729 systems, thus returning to full and verifiable compliance, Russia would bear sole responsibility for the end of the Treaty. NATO Allies also made clear that NATO would continue to closely review the security implications of Russian intermediate-range missiles and would continue to take steps necessary to ensure the credibility and effectiveness of the Alliance's overall deterrence and defence posture.

On 15 February 2019, NATO Secretary General Jens Stoltenberg recalled at the Munich Security Conference that it was on this very stage, at the Munich Security Conference in 2007, this was the place that President Putin first publically expressed his desire for Russia to leave the INF Treaty. A treaty that is only respected by one side will not keep us safe.

The Alliance did everything in its remit to encourage Russia to return to compliance before 2 August 2019 so as to preserve the INF Treaty.

On 26 June 2019, NATO Defence Ministers urged Russia once again to return to full and verifiable compliance. They also considered potential NATO measures (such as exercises, intelligence, surveillance and reconnaissance, air and missile defences, and conventional capabilities) and agreed that NATO would continue to ensure a safe, secure and effective nuclear deterrent. At the same time, Defence Ministers confirmed that NATO had no intention to deploy new land-based nuclear missiles in Europe, and did not want a new arms race.

On 2 August 2019, the United States' decision to withdraw from the Treaty took effect. NATO Allies issued a statement fully supporting the US decision, and attributing "sole responsibility" for the Treaty's demise to Russia. The statement made clear that NATO would respond in a "measured and responsible way" to the risks posed by Russia's SSC-8 system, with a "balanced, coordinated and defensive package of measures," ensuring credible and effective deterrence and defence. Allies also made clear their firm commitment to the preservation of effective international arms control, disarmament and non-proliferation.

Missiles banned under the INF Treaty:

Under the INF Treaty, the United States and Russia cannot possess, produce or flight-test a ground-launched cruise missile with a range capability of 500 to 5,500 kilometres, or possess or produce launchers of such missiles.

The INF Treaty gives precise definitions of the banned ground-launched ballistic and cruise missiles:

We will return to this topic later when we discuss the point on weapons of mass destruction.

# NATO'S CIVILIAN AND MILITARY STRUCTURES

In this section we will follow the structuring marked by the NATO texts and the philosophy that emerged from its complex institutional framework.

## Civilian organisation and structures

Initially based in London, the Headquarters was moved to Paris in 1952 before being transferred to Brussels, Belgium in 1967. NATO Headquarters, in Brussels, Belgium, is the political headquarters of the Alliance. It is home to national delegations of member countries and to liaison offices or diplomatic missions of Partner countries. The work of these delegations and missions is supported by the International Staff and the International Military Staff, which are also located within NATO Headquarters.

When the decisions taken by member countries have military implications, NATO has the military infrastructure and know-how in place to respond to demands. The Military Committee recommends measures considered necessary for the common defence of the Euro-Atlantic area and provides guidance to NATO's two strategic commanders (the Supreme Allied Commander Operations based in Mons, Belgium, and the Supreme Allied Commander Transformation in Norfolk, Virginia, United States). The Military Committee, located at NATO Headquarters, is supported by the International Military Staff, which plays a similar role to that of the International Staff for the North Atlantic Council.

Although their number varies with the passage of time, but we can say that there are approximately 4200 people working at NATO Headquarters on a full-time basis. Of these, some 2100 are members of the national delegations of member countries and staffs of national military representatives to NATO. There are approximately 1200 civilian members of the International Staff (including agencies and other NATO bodies) and just over 500 members of the International Military Staff. There are also just under 400 members of Partner missions to NATO. Civilian staff employed by NATO worldwide, including the staff of NATO agencies located outside Brussels and civilians serving on the staff of the military commands throughout NATO, number approximately 5200.

National delegations:

The national delegation of each member country has the status of an embassy and is headed by an ambassador (also referred to as a permanent representative), who acts on instructions from his/her capital and who reports back to the national authorities. The staff of the delegation comprises civil servants from the ministries of foreign affairs and other relevant ministries seconded to NATO to represent their respective countries.

The liaison offices of Partner countries are diplomatic missions headed by an ambassador or a head of mission who is responsible for communications between the national capital and NATO.

The Secretary General:

We will start by saying that the Secretary General has three main roles:

First and foremost, he is the chairman of the North Atlantic Council, the Defence Planning Committee and the Nuclear Planning Group as well as the chairman of the Euro-Atlantic Partnership Council, the NATO-Russia Council, the NATO-Ukraine Commission and the Mediterranean Cooperation Group.

Secondly, he is the principal spokesman of the Alliance and represents the Alliance in public on behalf of the member countries, reflecting their common positions on political issues.

Thirdly, he is the senior executive officer of the NATO International Staff, responsible for making appointments to the staff and overseeing its work.

The Secretary General is nominated by member governments for an initial period of four years. Usually an international statesman with ministerial experience in the government of one of the member countries, he acts as a decision facilitator, leading and guiding the process of consensus-building and decision-making throughout the Alliance. He may propose items for discussion and has the authority to use his good offices in cases of dispute between member countries.

His role allows him to exert considerable influence on the decision-making process while respecting the fundamental principle that the authority for taking decisions is invested only in the member governments themselves.

His influence is therefore exercised principally by encouraging and stimulating the member governments to take initiatives and, where necessary, to reconcile their positions in the interests of the Alliance as a whole.

As the Organization's senior representative, the Secretary General speaks on its behalf not only in public but also in its external relations with other organisations, with non-member country governments and with the international media.

The Secretary General is assisted by a Deputy Secretary General who replaces the Secretary General in his absence. The Deputy Secretary General is the chairman of a number of senior committees, ad hoc groups and working groups.

The International Staff:

The International Staff is an advisory and administrative body that supports the work of the national delegations at NATO Headquarters at different committee levels.

It follows up on the decisions of NATO committees and supports the process of consensus-building and decision-making. It is made up of personnel from the member countries of the Alliance recruited directly by NATO or seconded by their governments.


**The key functions of international staff**

In view of the changing security environment, NATO leaders are constantly reviewing the structure of the Organization's International Staff in order to reflect the Alliance's new missions and priorities. As a consequence, restructuring has become a permanent feature of the Organization.

Firstly, it is important to underline that the primary role of the International Staff is to provide advice, guidance and administrative support to the national delegations at NATO Headquarters. Secondly, from a purely organisational point of view, it must be noted that all divisions are headed by an Assistant Secretary General, who is supported by one or two Deputy Assistant Secretary Generals, and the independent offices are headed by directors.

The Secretary General, who heads the International Staff (IS) but is also from an administrative point of view a member of the IS, has a Private Office that includes a director and staff, the Deputy Secretary General, the Office of the Legal Adviser and a Policy Planning Unit.

A) Providing political advice and policy guidance:

Let us mention that the political aspects of NATO's fundamental security tasks need to be managed on a daily basis. They embrace a wide range of issues at the top of the Alliance's political agenda, which include regional, economic and security affairs, relations with other international organisations and relations with Partner countries.

A number of high-level bodies need to be informed on these political matters and advised on current and future policy issues. The North Atlantic Council, for instance, and other NATO committees can request information, while other sections of the International Staff and the International Military Staff need to be advised on current and future policy issues. The Secretary General also requests input relevant to NATO's political agenda such as background notes, up-to-date reports, and speeches. In addition, for meetings involving NATO and Partner countries, as well as for the political contacts with the respective national authorities, political preparation is also necessary to support the political consultation process. This is also provided by the staff responsible for political matters.

In summary, the aim is to provide political guidance for the implementation of the policy areas listed above. In relation to the enlargement process, for instance, advice, support and assistance to member countries, invited countries and relevant NATO bodies in handling the process of accession of new member countries is provided. The same applies, for instance, to the continuity of the Membership Action Plan, the development of the NATO-EU strategic partnership and the expansion of cooperation with Partner countries.

Other matters addressed include the provision of political country area expertise and support for operational matters in the crisis management field; the coordination of political and economic aspects of cooperation in relation to NATO's role in the fight against terrorism; and the coordination of political aspects involved in the enhancement of the readiness and effectiveness of Allied forces for operations aimed at responding to the use of weapons of mass destruction.

Contributions are also made to the public relations activities of the Alliance designed to inform external audiences in member and Partner countries as well as elsewhere about NATO's tasks, policies and objectives.

B) Developing and implementing the defence policy and planning dimension:

Developing and implementing the defence policy aspects of NATO's fundamental security tasks includes defence planning, nuclear policy and defence against weapons of mass destruction. Moreover, the defence dimension also comprises operational issues.

Trying to be as accurate as possible, we will say that just as the Secretary General, member countries and Partner countries need political advice, they also need support in developing and implementing the defence policy and planning dimension of Alliance and partnership activities.

This includes the Alliance's response to terrorism; the defence perspective of NATO's cooperation with the European Union (including the Berlin Plus arrangements), the United Nations and other international organisations; and politico-military aspects of NATO's transformation agenda and capabilities initiatives, including NATO's command structure and force structure (and in particular the implementation of the NATO Response Force), the Prague Capabilities Commitment and policy guidance for capabilities development.

It also comprises support on the Planning and Review Process for Partner countries, as well as other defence aspects of cooperation with countries within the Partnership for Peace, the Mediterranean Dialogue and other countries as required.

Besides, the development and implementation of logistics policy and planning initiatives within NATO and between

NATO Headquarters, external NATO bodies, NATO's Strategic Commands and the member countries also comes under defence policy and planning, as well as the review of NATO's nuclear policy guidance and force posture, promoting public understanding of the nuclear elements of NATO strategy, and the training and exercising of nuclear consultation procedures.

In addition, the Weapons of Mass Destruction Centre (WMDC), located at NATO Headquarters, is responsible for supporting the sharing of information and intelligence in this field.
Managing.

C) Managing NATO's operational commitments and crisis response capabilities:

In this point, overseeing the operational capability required to meet NATO's deterrence, defence and crisis management tasks is essential for the success of NATO missions. Responsibilities include NATO's crisis management and peacekeeping activities and civil emergency planning and exercises, which encompass NATO's operational commitments.

For it the International Staff supports and advises the senior committees involved in the above areas and prepares and follows up their discussions and decisions. This includes the Policy Coordination Group, the Senior Civil Emergency Planning Committee, and the Council Operations and Exercises Committee. It also contributes to the implementation of the NATO-Russia Work Programme, the NATO-Ukraine Action Plan and the Mediterranean

Dialogue Work Programme in each of these specific areas of responsibility.

Let us mention that the information and communications processes and technological aspects of crisis management mechanisms, joint exercises and civil emergency planning are also managed in liaison with other NATO bodies and other international organisations.

In the operations sphere, there are Operational Task Forces to oversee the role of NATO-led forces in different crisis areas. Direction is provided for the future elaboration of the Alliance's crisis management procedures and arrangements, and the Situation Centre ensures continuous and secure links between NATO Headquarters and NATO capitals, Strategic Commands, other military structures and other organisations.

Finally, in the civil emergency planning field, the International Staff supports the work of specialised civil emergency planning boards and committees responsible for drawing up arrangements relating to the use of civil resources in support of NATO operations and the protection of the civilian population.

It also maintains contacts and consultation with the United Nations, the World Health Organization, the International Atomic Energy Agency and the Organization for the Prohibition of Chemical Weapons, as well as the appropriate bodies in the framework of joint NATO-EU activities relating to civil emergencies.

The Euro-Atlantic Disaster Response Coordination Centre (EADRCC) coordinates disaster assistance for EAPC countries and is responsible for maintaining and updating EADRCC organisation and procedures for responding to emergencies.

D) Developing assets and capabilities:

Another key area of responsibility of the International Staff is the development of and investment in assets and capabilities aimed at enhancing the Alliance's defence capacity, including armaments planning, air defence and security investment.

Thus, policy, technical, financial and procedural expertise relating to armaments, air defence, airspace management and security investment is provided, and work is undertaken for the development of military capabilities and overseeing investment in NATO common-funded assets to ensure that forces assigned to the Alliance are properly equipped, interoperable and able to undertake the full range of military missions.

Work is also conducted in developing cooperation with partner countries in the context of the Partnership for Peace, the Mediterranean Dialogue and the Istanbul Cooperation Initiative, as well as the special relationships with Russia and Ukraine.

The work is divided into three main areas: armaments, air defence and airspace management, and security investment.

- Armaments provides support for the work of the Conference of National Armaments Directors (CNAD) and its subordinate structures, focusing on the collaborative development and acquisition of military equipment. It comprises, for instance, units for land, air and naval armaments, as well as a dedicated Counter-Terrorism Technology Unit.

- In the area of air defence and airspace management, policy and technical expertise is provided, as is support for two senior NATO committees: the NATO Air Defence Committee, which harmonises national air defence policies and programmes, and the NATO Air Traffic Management Committee, which develops harmonised civil-military policy guidance and requirements on the use of airspace in support of Alliance tasks and missions.

- With regard to security investment, the key objective is to ensure the timely provision of common-funded capital investments in support of NATO's operational requirements. Funding for these capabilities is provided through the NATO Security Investment Programme (NSIP). Support is also given to the following committees: the Senior Resource Board (SRB), Military Budget Committee, the Infrastructure Committee, the NATO Consultation, Command and Control (C3) Board, and the Missile Defence Project Group.

The area of assets and capabilities also encompasses the International Staff support element for the NATO Headquarters Consultation, Command and Control Staff (NHQC3S), the Office of the Chairman of the Senior Resource Board and a Resource Policy Coordination Section.

E) Communicating with the wider public:

Obviously, the Organization has an obligation to inform the wider public in member and Partner countries about NATO's activities and policies. It does this through a variety of communication activities including contacts with the media, the NATO website, print and electronic publications, and seminars and conferences.

These efforts contribute to raising public awareness and knowledge of the issues with which NATO is concerned, and help to promote constructive debate about NATO policies and objectives.

That is why the staff working in press and public relations constitute one of NATO's principal public interfaces with external audiences worldwide. They provide support for the NATO Secretary General in his role as principal spokesman for the Alliance and arrange briefings and interviews with journalists, press conferences, press tours, media monitoring, audio-visual media support and exhibits.

Cooperation programmes in NATO member and Partner countries are organised and visits, seminars and conferences involving opinion leaders, parliamentarians, civic society groupings and experts in different fields are held in different countries. Grants and other forms of support for special projects are made available, and print and electronic publications are distributed on request. Publications work in print and electronic formats covers a broad range of NATO-related topics, is often produced in a variety of NATO and Partner country languages, and is disseminated worldwide.

NATO's website provides access to up-to-date information on NATO policies and activities including public statements, background information, and official documents, as well as video interviews, audio files, real-time coverage of major NATO-related events and the resources of the NATO media library.

Staff also work closely with the Public Information Adviser to the Chairman of the Military Committee and assist in coordinating public diplomacy activities in other parts of the International Staff and entities within the NATO structure.

F) Cooperating with the science community:

Since the attacks on the United States of 11 September 2001 and the increased focus of NATO policies on new challenges and threats, including terrorism, NATO's science programme has directed increasing support to collaborative research projects related to defence against terrorism and other threats to security.

The programme serves to strengthen cooperation between NATO and Partner countries through different forms of support mechanisms including collaborative grants in priority areas, currently defined as defence against terrorism, countering other threats to security, and Partner-country priorities. Collaboration is between research scientists in NATO member countries and those in eligible Partner or Mediterranean Dialogue countries. Computer networking support for Partner countries, particularly in Central Asia and the Caucasus, is also an important element of the activities undertaken by the programme.

The programme also deals with two other research areas: human and societal dynamics and security-related civil science. Studies and workshops are organised, bringing together experts from different government agencies in NATO member countries, Partner countries and Mediterranean Dialogue countries in order to concentrate inter-governmental action on pressing areas of environmental security.

These areas include the environmental impact of military activities, regional studies including cross-border activities, the prevention of conflicts arising in relation to scarcity of resources, emerging risks to the environment and society with the potential to cause economic cultural or political instability, and non-traditional threats to security. Examples of areas of study in the latter sphere include food chain security, risk response strategies and security of waterways, ports and harbours.

G) Managing staff, finances and security standards:

Logically, the effective management of the International Staff and the actual running of the Headquarters require support and conference services, information, human and financial management, as well as the support of security services.

As usual, one of the main functions is to handle all human resources management and development matters for the International Staff including administration, recruitment, contracts, training and development, performance management, personnel support, health and social matters, civilian personnel policies and compensation and benefits. An underlying principle applied throughout is the improvement of the gender balance and diversity of the staff working at NATO Headquarters.

In case it is of interest we will say that it is also the responsibility of the International Staff working in Headquarters management to oversee the NATO Staff Centre, which comprises catering, sport and recreational facilities, and to stay in regular contact with NATO Staff Association representatives. The support services provided by Headquarters management include conference services (interpretation and translation), and an information and systems management service. These support services also include building services and teams responsible for the interior and exterior utilities, maintenance and transport.

NATO's civil budget is operated on an objective-based budgeting (OBB) system. In conjunction with the Civil Budget Committee, staff are responsible for managing the annual budget preparation and approval cycle based on political guidance issued by the North Atlantic Council. They also monitor and manage the procurement cycle from contracting and purchasing to reception, inventories and distribution.

You have to know that the Financial Controller is appointed by the North Atlantic Council and is responsible for the call-up of funds from the member countries and for the control of expenditures in accordance with NATO's financial regulations, within the framework of relevant budgets, namely the civil budget, the budget for the new NATO headquarters and the pension budget. Audits of the management process are carried out in order to identify the principal management risks and keep them under control.

Besides, there is an independent International Board of Auditors for NATO, composed of government officials from the auditing bodies of member countries appointed by and responsible to the North Atlantic Council for auditing the financial accounts of the various NATO bodies under their respective budgets. Its principal task is to provide the Council and member governments with the assurance that common funds are properly used for the settlement of authorised expenditure and that expenditure is within the physical and financial authorisations granted.

Organisational, procedural and administrative support is also provided to all official meetings at ambassadorial, ministerial and summit level, and to special events at NATO Headquarters and abroad. The same staff works closely with the Private Office of the Secretary General, who has overall responsibility for the running of the Organization.

In addition, there is an independent office in the International Staff that is responsible for ensuring the coordination and implementation of security standards throughout NATO. It carries out periodic surveys of security systems and is responsible for security at NATO Headquarters in Brussels. Moreover, it is responsible for the overall coordination of security for NATO among member countries, Partner countries and Mediterranean Dialogue countries and NATO civil and military bodies, for the correct implementation of NATO security policy throughout the Alliance, and for the evaluation and implementation of countermeasures against terrorist and intelligence threats.

The office has three main functions: policy oversight, security intelligence and protective security.

I.    With regard to policy oversight, inspections and surveys mandated by the North Atlantic Council in NATO member countries, NATO civil and military bodies and in Partner and Mediterranean Dialogue countries are carried out, and the proper protection of NATO information is certified. Security policy, directives, guidance and supporting documents are developed and revised for consideration by the NATO Security Committee and, where required, by the North Atlantic Council. Security agreements with non-NATO countries and international organisations receiving NATO classified material in support of cooperation activities approved by the North Atlantic Council are negotiated. Moreover, security accreditation of communication and information systems in NATO civil bodies is undertaken and advice given on information security aspects related to NATO multinational communication and information systems. The application of security risk management procedures also needs to be ensured in NATO civil bodies, as does the coordination of the NATO elements of an associated security risk assessment methodology.

II.   Security intelligence deals with counter-intelligence policy and oversight throughout NATO, providing threat-related information to the North Atlantic Council and the other principal decision-making bodies in NATO as well as to the NATO Military Committee, through a Terrorist Threat Intelligence Unit. Staff working in this area coordinate the work of and provide the secretariat for the NATO Special Committee. They also support and, in limited cases, conduct special inquiries and espionage investigations, manage all matters relating to personnel security at NATO Headquarters and in NATO civil and military bodies, and manage the Secretary General's Close Protection Unit. Protective

III. Protective security encompasses a number of elements: coordinating protective security programmes and operations, including physical, personnel and information security measures at NATO Headquarters, providing advice on protective security measures for the new NATO headquarters buildings, coordinating the security aspects of NATO ministerial and other high-level meetings at NATO Headquarters and in NATO member and Partner countries, and under NATO's Cyber Defence Programme, managing and operating the NATO Computer Incident Response Capability Coordination Centre (NCIRC CC), responsible for the coordination throughout NATO of information security awareness and responses to computer security incidents such as computer virus outbreaks and network attacks.

**Military organisation and structures**

This section describes the military components of the Organization, which are the Military Committee (NATO's senior military authority), the two Strategic Commanders and the military command structure. The work of the Military Committee is supported by the International Military Staff.

The Military Committee:

The Military Committee (MC) is the senior military authority in NATO under the overall political authority of the North Atlantic Council and, as appropriate, of the Defence Planning Committee and the Nuclear Planning Group. It meets under the chairmanship of an elected chairman (CMC) and is the primary source of military advice to the North Atlantic Council, Defence Planning Committee and Nuclear Planning Group. Its members are senior military officers who serve as national military representatives (MILREPs) in permanent session, representing their chiefs of defence (CHODs).

A civilian official represents Iceland, which has no military forces. The Military Committee also meets regularly at a higher level, namely at the level of chiefs of defence, when the two NATO Strategic Commanders are invited to attend.

Thus, on a day-to-day basis, the military representatives work in a national capacity, representing the best interests of their countries while remaining open to negotiation and discussion so that consensus can be reached. This often involves reaching agreement on acceptable compromises, when this is in the interests of the Alliance as a whole and serves to advance its overall objectives and policy goals. The military representatives therefore have adequate authority to enable the Military Committee to discharge its collective tasks and to reach prompt decisions.

It is essential to know that the Committee is responsible for recommending to NATO's political authorities those measures considered necessary for the common defence of the NATO area and for the implementation of operational decisions taken by the North Atlantic Council. Its principal role is to provide direction and advice on military policy and strategy. It provides guidance on military matters to the NATO Strategic Commanders, whose representatives attend its meetings, and is responsible for the overall conduct of the military affairs of the Alliance under the authority of the Council, as well as for the efficient operation of Military Committee agencies.

Equally, the Committee assists in developing overall strategic concepts for the Alliance and prepares an annual long-term assessment of the strength and capabilities of countries and areas posing a risk to NATO's interests.

In times of crisis, tension or war, and in relation to military operations undertaken by the Alliance (such as those in Kosovo and Afghanistan), its role is to advise the Council or the Defence Planning Committee of the military situation and its implications, and to make recommendations on the use of military force, the implementation of contingency plans and the development of appropriate rules of engagement.

We can say that the Military Committee normally meets every Thursday, following the regular Wednesday meeting of the Council, so that it can follow up promptly on Council decisions. In practice, meetings are convened whenever necessary, and both the Council and the Military Committee normally meet much more often.

As a result of the Alliance's role in Kosovo and Afghanistan and its supporting role in relation to Iraq and Sudan, there is a heightened need for the Council and Military Committee to meet more frequently to discuss operational matters.

Indeed, as a result of the internal and external transformation of Alliance structures, the intensification of partnership and cooperation with other countries, the creation of new institutions to oversee these developments, and in particular the emergence of new threats and the development of the Alliance's role in combating terrorism, the frequency of meetings of all the decision-making bodies of the Alliance has greatly increased in recent years.

The Military Committee in Chiefs of Staff Session normally meets three times a year. Two of these Military Committee meetings are held in Brussels and one is hosted by NATO member countries, on a rotational basis.

In the framework of the Euro-Atlantic Partnership Council (EAPC) and Partnership for Peace (PfP), the Military Committee meets regularly with EAPC/PfP countries at the level of national military representatives (once a month) and at the level of chiefs of defence (twice a year) to discuss military cooperation issues.

The Military Committee also meets in different formats in the framework of the NATO-Russia Council and the NATO-Ukraine Commission.

Alike, since January 2001, the Military Committee of NATO has met regularly with the Military Committee of the European Union on issues of common interest relating to security, defence and crisis management.

The first meeting of the Military Committee in Chiefs of Staff session with the participation of the chiefs of defence of the Mediterranean Dialogue countries was held in November 2004.

The Chairman of the Military Committee:

We have to know that the Chairman of the Military Committee is nominated by the chiefs of defence and appointed for a three-year term of office. He acts in an international capacity and his authority stems from the Military Committee, to which he is responsible in the performance of his duties. He normally chairs all meetings of the Military Committee. In his absence, the Deputy Chairman of the Military Committee (DCMC) takes the chair.

The Chairman of the Military Committee is both its spokesman and representative. He directs its day-to-day business and acts on behalf of the Committee in issuing the necessary directives and guidance to the Director of the International Military Staff. He represents the Military Committee at high-level meetings, such as those of the North Atlantic Council, the Defence Planning Committee and the Nuclear Planning Group, providing advice on military matters when required.

We can assume that by virtue of his appointment, the Chairman of the Committee also has an important public role and is the senior military spokesman for the Alliance in contacts with the press and media.

He undertakes official visits and representational duties on behalf of the Committee, both in NATO countries and in countries with which NATO is developing closer contacts in the framework of the Partnership for Peace programme, the Euro-Atlantic Partnership Council, the NATO-Russia Council, the NATO-Ukraine Commission, and the Mediterranean Dialogue. The Chairman is also the ex-officio chairman of the NATO Defence College Academic Advisory Board.

NATO's strategic commanders

The Strategic Commanders are: the Supreme Allied Commander Europe (SACEUR) and the Supreme Allied Commander Transformation (SACT). They are responsible to the Military Committee for the overall direction and conduct of all Alliance military matters within their areas of responsibility. They also provide advice to the Military Committee on their command responsibilities.

They normally attend the Military Committee meeting in Chiefs of Staff Session but may be called upon to brief the Military Committee in Permanent Session when required. For day-to-day business, each has a representative at NATO Headquarters of general or flag officer rank who assists them by maintaining close links with both the political and military staffs within the Headquarters and ensuring that the flow of information and communications in both directions works efficiently. These representatives attend meetings of the Military Committee in Permanent Session and provide advice on Military Committee business relating to their respective Commands.

The Military Command Structure:

The Military Command Structure of NATO, as distinct from the NATO Force Structure, is the mechanism which enables NATO's military authorities to command and control the forces assigned to them for joint operations involving more than one service branch, army, navy, air force. It is based on a hierarchical structure of Strategic Commands and Subordinate Commands.

We must clarify that the NATO Force Structure consists of the organisational arrangements that bring together the forces placed at the Alliance's disposal by the member nations, temporarily or permanently, along with their associated command and control structures, either as part of NATO's multinational forces or as additional national contributions to NATO. These forces are available for NATO operations in accordance with predetermined readiness criteria.

Changes to the NATO Force Structure introduced over recent years have placed the emphasis on smaller, more mobile forces that can be used flexibly for a range of military tasks, as opposed to the large, heavily armed concentrations of forces in permanent fixed headquarters that were a feature of Cold War force structures. While the latter were equipped and trained for major defence operations against an invading army, the majority of forces that comprise NATO's present-day force structure are designed to be moved rapidly to the area of crisis or conflict where they are required and to have the capability to fulfil their role away from their home bases.

Let us underline that the above changes to the force structure have brought about a parallel need for changes to the NATO Command Structure. These have concentrated on reductions in the number of commands within the structure and on rationalisation of the system of command and control linking the different elements which together make up NATO's military capabilities. These changes are designed to permit NATO's Strategic Commanders to exercise more effective command and control of the forces assigned to them, drawing on the full range of military capabilities needed to undertake the kind of operations that may be assigned to them in today's vastly different security environment.

Important to know that the present-day NATO Command Structure reflects changing strategic circumstances attributable to a number of factors including the accession of new member countries, NATO's evolving strategic partnership with the European Union, its cooperation with Partner countries and relations with other non-NATO countries, new security challenges including the evolving threat of terrorism, and the proliferation of weapons of mass destruction. The command structure is designed to cope with the likely tasks, risks and potential threats facing the Alliance across the board and to meet them when and where it may be required by the North Atlantic Council to do so.

At the centre of the command structure are two strategic commands:

a) Of these, one is focused on planning and executing all the operations that the North Atlantic Council has agreed to undertake.

b) The other is concerned with the transformation of NATO's military capabilities to meet changing requirements and enable the military forces made available to the Alliance to carry out the full range of military tasks entrusted to them. The transformation process is a continuous one. It calls for the proactive development and integration of innovative concepts, doctrines and capabilities designed to improve the effectiveness and interoperability of the forces that NATO and Partner countries may make available for NATO-led military operations.

There are also a number of subordinate military headquarters and other components of the command structure located in different NATO member countries.

With the separation of strategic command responsibilities along operational and functional lines, all the operational responsibilities formerly shared by Allied Command Europe and Allied Command Atlantic are now vested in a single European-based Strategic Command called Allied Command Operations (ACO), in Mons, Belgium, under the responsibility of the Supreme Allied Commander Europe (SACEUR). The appointed officer is "dual-hatted" and serves simultaneously as the Commander of the United States European Command.

In the same way, the second strategic-level command, known as Allied Command Transformation (ACT), is based in the United States and comes under the responsibility of the Supreme Allied Commander Transformation (SACT), who serves simultaneously as the Commander of the United States Joint Forces Command. This helps to maintain a strong transatlantic link and at the same time ensures access for NATO's forces to the transformational process being undertaken by the United States in relation to its national military forces.

Both NATO's Strategic Commanders carry out roles and missions assigned to them by the North Atlantic Council or in some circumstances by NATO's Defence Planning Committee, under the direction of the Military Committee.

Their responsibilities and tasks are based on the objectives Ooutlined in the Alliance's Strategic Concept and in relevant Military Committee documents. In broad terms they are as follows:

A) The Supreme Allied Commander, Europe (SACEUR):

We will start by saying that the Supreme Allied Commander Europe (SACEUR) for Allied Command Operations (ACO) is task with contributing to the peace, security and territorial integrity of Alliance member countries by assessing risks and threats, conducting military planning, and identifying and requesting the forces needed to undertake the full range of Alliance missions, as and when agreed upon by the North Atlantic Council and wherever they might be required.

Thus, SACEUR contributes to the Alliance's crisis management arrangements and provides for the effective defence of the territory of NATO countries and of their forces. If aggression occurs, or if the North Atlantic Council believes that aggression is imminent, SACEUR executes all the military measures within the authority and capabilities of his Command needed to demonstrate Allied solidarity and preparedness to maintain the integrity of Alliance territory, to safeguard the freedom of the seas and lines of communication and trade, and to preserve the security of NATO member countries or restore it if it has been infringed.

Allied Command Operations also contributes to the process of ensuring that the forces that make up the NATO Force Structure are provided both now and in the future with effective combined or joint military headquarters able to call on the military capabilities needed to perform their tasks. It does so in consultation with Allied Command Transformation by synchronising operational activities and elements of the command structure that have an operational role in the Alliance's transformation efforts.

Other tasks that are also under the responsibility of the Supreme Allied Commander Europe include:

- contributing to stability throughout the Euro-Atlantic area by developing and participating in military-to-military contacts and other cooperation activities and exercises undertaken in the framework of the Partnership for Peace (PfP) and of activities undertaken to enhance NATO's relationships with Russia, Ukraine and Mediterranean Dialogue countries;

- conducting analysis at the strategic level to identify capability shortfalls and to assign priorities to them;

-

- managing the resources allocated by NATO for operations and exercises, and accomplishing the operational missions and tasks assigned by the North Atlantic Council; and

- in conjunction with Allied Command Transformation, developing and conducting training programmes and exercises in combined and joint procedures for the

military headquarters and forces of NATO
and Partner countries.

B) The Supreme Allied Commander Transformation
   (SACT):

Allied Command Transformation (ACT) comes under the
authority of the Supreme Allied Commander
Transformation (SACT), whose responsibilities can be
summarised as follows:

- contributing to the preservation of peace and
  security and of the territorial integrity of
  Alliance member countries by assuming the
  lead role at the strategic command level in
  the transformation of NATO's military
  structures, forces, capabilities and doctrines
  in order to improve the military effectiveness
  of the Alliance;

- conducting operational analysis at the
  strategic level, in cooperation with ACO, in
  order to identify and prioritise the type and
  scale of future capability and interoperability
  requirements and to channel the results into
  NATO's overall defence planning process;

- integrating and synchronising NATO's
  transformation efforts, in cooperation with
  ACO, with the operational activities and
  other elements of the command structure in
  order to contribute to the process of ensuring
  that NATO forces are provided both now and
  in the future with effective combined or joint

military headquarters able to call on the military capabilities needed to perform their tasks;

- exploring concepts and promoting doctrine development,

- conducting experiments and supporting the research and acquisition processes involved in the development of new technologies; in fulfilling this function, interacting with appropriate NATO agencies and project management boards in order to identify opportunities for improved interoperability and standardisation and to deliver qualitatively transformed capabilities for the benefit of the Alliance;

- managing commonly funded resources allocated for NATO's transformation programmes in order to provide timely, cost-effective solutions for operational requirements;

- conducting training and education programmes in order to provide the Alliance with leaders, specialists and headquarters staffs trained to common NATO standards and capable of operating effectively in a combined and joint force military environment;

- establishing and maintaining procedures designed to ensure the continuous adaptation of the organisations, concepts, resources and education programmes required to promote NATO's transformation efforts;

- supporting the exercise requirements of Allied Command Operations throughout their planning, execution and assessment phases.

C) Shared roles and responsibilities:

It should be clarified that while the new command structure provides for close cooperation between the two strategic commands, there is a clear division of their respective responsibilities in order to avoid unnecessary overlap and duplication of effort. Each command therefore has well-identified lead and supporting roles. In specific areas where a strategic command has the lead role, it is responsible for providing the formal input to the Military Committee but receives assistance from the other strategic command as required. Conversely, where the strategic command is in the supporting role, it is responsible for channelling command-level and staff-level advice to the strategic command in the leading role.

This way, while both strategic commands have responsibilities in NATO's defence planning process, for the purposes of ensuring a streamlined and coherent process, the lead role is undertaken by Allied Command Transformation for military aspects such as reviewing defence requirements, force planning, armaments and logistics planning, and command and control planning, as well as for the Partnership for Peace (PfP) Planning and Review Process. Allied Command Operations has the lead on military aspects of civil emergency planning and nuclear planning and for suitability and risk assessments in force planning.

According to the philosophy of the Alliance, command and control of forces, including operational planning, is the preserve of SACEUR. Where joint and combined concepts and doctrine are concerned, on the other hand, including PfP military concepts, the responsibility lies with SACT, drawing on lessons learned from operations and exercises as a basis for introducing changes relating to concepts, doctrine and capabilities. These then form the basis of strategic directives and procedures for operations developed by Allied Command Operations.

Thus, each of the strategic commands has individual management structures and budgets and individual responsibilities relating to them, including strategic management, financial planning and resource management.

The management of capability packages is another responsibility shared by both Strategic Commands. SACT focuses on the development of capabilities to improve joint and combined effectiveness for the full range of Alliance missions. SACEUR is responsible for the development of capabilities required for the conduct of operations.

In the intelligence field, Allied Command Operations provides intelligence support for operational planning and operations, while Allied Command Transformation concentrates on the long-term analysis of trends and development of intelligence concepts and capabilities. Similarly, in the sphere of communications and information systems, the division of responsibility enables ACO to focus on operational planning and identifying shortfalls, while

ACT concentrates on future concepts, capabilities and structures.

And with regard to exercises, training, evaluation and experimentation, SACT has the leading role for NATO and PfP joint individual education and training and associated policy. The design, conduct and assessment of experiments to assist in the development and testing of emerging concepts, doctrine and technology form part of this task. Close coordination with SACEUR and the member countries takes place with regard to scheduling and access to forces for training, exercises and experimentation undertaken for the purposes of fulfilling transformation objectives.

NATO and PfP collective training of assigned forces and subordinate elements of the operational command structure are the responsibility of SACEUR supported by ACT, which provides exercise design, planning and evaluation assistance.

Finally, Allied Command Transformation has the leading role in scientific research and development, although Allied Command Operations conducts its own operational analysis and provides technical support for the command structure and for operations.

  D) The operational structure:

It is important to know that all NATO operations draw on deployable or static elements and capabilities available to the integrated command structure and force structure, tailored to the requirements and challenges of the specific operation. This applies whether they are operations undertaken by the Alliance in response to a threat to one or more of the member countries in accordance with Article 5 of the North Atlantic Treaty (known as Article 5 operations), or peace support or other military operations decided upon by the North Atlantic Council (non-Article 5 operations).

The command and control structure functions at three levels: the strategic, operational and component levels.

> 1-. At the strategic level, Allied Joint Forces are employed within a political-military framework endorsed by the Military Committee and approved by the North Atlantic Council, designed to fulfil the strategic objectives of the Alliance. Overall command of any operation, at the strategic level, is assumed by SACEUR, who exercises this responsibility from the headquarters of Allied Command Operations at the Supreme Headquarters Allied Powers in Europe (SHAPE) in Mons, Belgium. SACEUR is responsible for the preparation and conduct of all Alliance military operations, in accordance with the division of responsibilities between the strategic commands outlined above. He issues strategic military direction to the subordinate commanders and coordinates the multinational support, reinforcement and designation of the different elements and components of the command structure.
>
> Otherwise, in circumstances where political decisions have been taken based on the framework agreement reached between NATO and the European Union regarding NATO support for EU-

led military operations, SACEUR may also be required to provide a headquarters capability for such operations, open to the participation of all Allies.

Similarly, the involvement of Partner countries in command arrangements is set out in an agreed Political-Military Framework for NATO-led PfP operations.

➢ 2-. At the operational level, the planning and conduct of operations, based on the strategic military guidance received, is in the hands of the designated operational-level commander who exercises his responsibilities through a joint permanent or deployable operational headquarters.

The Alliance has three operational-level standing joint headquarters: two Joint Force Command (JFC) Headquarters based in Brunssum, the Netherlands, and Naples, Italy, and a third, more limited Joint Headquarters based in Lisbon, Portugal. Differing arrangements are in place to ensure that, if necessary, at least two operations can be conducted concurrently, that they can be sustained for the period of time required and that their varying components can be relieved. Operational command and control of the NATO Response Force rotates between the three Joint Headquarters.

The two Joint Force Commands have subordinate land, maritime and air Component Commands. The third Joint Headquarters has no permanent operational command responsibilities and is primarily responsible for providing support for Combined Joint Task Force operations.

> 3-. At the Component Command level, a number of Component Command Headquarters provide service-specific expertise for Joint Force commanders at the operational level, as well as advice on joint operational planning and execution. There are two Component Command-level land headquarters in Germany and Spain respectively, namely CC-Land Headquarters, Heidelberg, and CC-Land Headquarters, Madrid.

> There are also two static Component Command air headquarters in Ramstein, Germany and in Izmir, Turkey, and two static Component Command-Maritime headquarters in Northwood, England and in Naples, Italy. Each of the Component Command headquarters is supported by other specialised entities and subordinate elements, depending on the nature and scale of the operations involved, and can be augmented if necessary by additional elements and personnel at appropriate levels of readiness and training.

In addition to the above structure of Component Commands, an Allied Submarine Command Headquarters located in Norfolk, Virginia, subordinate to SACEUR but funded nationally by the United States, has the lead responsibility for the overall coordination of Alliance submarine matters in conjunction with the Component Command-Maritime Headquarters in Northwood and in Naples.

E) The transformational process and structure:

Let us point out that the transformation of NATO's military capabilities comprises closely linked functions working together to address the critical task of introducing the improvements in capabilities needed to ensure that the Alliance can meet the challenges of the current and future security environment.

These functions include, for example, assessing the future operating environment, identifying future strategic-level joint concepts, developing, integrating and testing new operational concepts, doctrines, organisational structures, capabilities and technologies, and contributing to the implementation of improvements.

The Organization of Allied Command Transformation is designed to enable the various processes involved in bringing about improvements to be undertaken within a coherent, integrated transformation framework. The work involves several distinct but related functions, including strategic concepts, policy and requirements identification; capability planning and implementation; joint and combined concept development, experimentation, assessment and doctrine; future capabilities research and technology; and training and education.

The headquarters of SACT, in Norfolk, Virginia, maintains close links to the United States Joint Forces Command and is the focal point for overseeing the entire transformation process and coordinating the above functions. It does so by means of a network of centres and entities within NATO's military structure located in Europe and North America, contributing to different aspects of the transformation process. Some of these centres or entities support more than one function or process.

Thus, in organisational terms, in addition to its headquarters in Norfolk, Allied Command Transformation has an ACT Staff Element based at NATO Headquarters in Brussels to support SACT's representative to the Military Committee, plus an ACT Staff Element co-located with SHAPE, thereby ensuring liaison and coordination with Allied Command Operations, the International Staff, the International Military Staff, national military representatives and other NATO bodies and agencies responsible for defence and resource planning and implementation issues.

Within the ACT structure there is also a Europe-based Joint Warfare Centre (JWC) located in Stavanger, Norway, which promotes and conducts NATO's joint and combined experimentation, analysis, and doctrine development processes to maximise transformational synergy and to improve NATO's capabilities and interoperability. The Centre contributes to developmental work on new technologies, modelling and simulation and conducts training and development programmes for new concepts and doctrine.

Then, the ACT structure includes a Joint Analysis and Lessons Learned Centre located in Monsanto, Portugal, which feeds the results of joint analysis work and lessons learned back into the transformation network.

Through its Joint Force Training Centre located in Bydgoszcz, Poland, the JWC provides training programmes to assist both strategic commands and contributes to joint force training evaluation undertaken by Allied Command

Operations. It maintains formal links with other NATO agencies and bodies and with national and multinational training centres and facilities.

The Undersea Research Centre at La Spezia, Italy, conducts research and integrates national efforts that support NATO's undersea operational and transformational requirements. A NATO Maritime Interdiction Operational Training Centre is in the process of being set up in Souda Bay, Greece.

And there are also a number of educational facilities within the NATO military structure, including the NATO Defence College in Rome, Italy, the NATO School in Oberammergau, Germany, and the NATO CIS School in Latina, Italy, which coordinate their activities with Allied Command Transformation.

Besides, a number of Centres of Excellence, funded nationally or multi-nationally, provide opportunities for improving interoperability and capabilities, testing and developing doctrine, and validating new concepts through experimentation. Cooperation between the Centres of Excellence and the Strategic Commanders is established on the basis of specific memoranda of understanding drawn up between the participating nations and the appropriate Strategic Command.

F) Military forces:

As we know, in general, NATO does not have independent military forces, other than those contributed by the member countries to military operations. Therefore, when the North Atlantic Council decides to launch an operation, forces have to be made available by member countries through a force generation process. This may include forces of non-NATO member countries, such as Partnership for Peace (PfP) and Mediterranean Dialogue countries. Once these forces have completed their mission, they are reintegrated into their national military structures.

Working together requires a compatibility of equipment and a sufficient level of interoperability, for example to enable forces to be refuelled or resupplied by another country. The forces of NATO and Partner countries also train together, participate in courses on standardised operational procedures and language, conduct military simulations and take part in other multinational exercises, all of which serve to enhance their ability to undertake combined (multinational) operations.

However, there are a number of common defence capabilities, most significantly a fleet of Airborne Warning and Control Systems (AWACS) aircraft, which provide air surveillance, early warning, and command and control.

A milestone in this story was that following the terrorist attacks on the United States in 2001, AWACS aircraft were sent to patrol United States territory. At the request of Turkey in 2003, they were sent to guard Turkish territory against the possibility of an attack arising from the Iraqi conflict. They also frequently contribute to the security of major events, such as the 2004 Olympic Games in Athens.

In addition, AWACS aircraft, the NATO Airborne Early Warning and Control capability includes a number of multinational support and training aircraft, an integrated radar system and shared infrastructure installations. Such forces are manned by multinational staff provided by the member countries for these specific roles in the framework of the Alliance's integrated force structure.

G) Civil-military cooperation:

We can show that in the present-day security environment, close co-operation between civil and military bodies has become an increasingly important factor in the successful conduct of military operations, especially in post-conflict peace-support operations such as those undertaken by the Alliance in the Balkans and in Afghanistan. Civil-military cooperation encompasses the vital coordination needed between NATO operational commands and civilian organisations, including local authorities and the local population in any given area of operations, as well as international, national and non-governmental organisations and agencies.

A multinational project sponsored by six NATO countries (Czech Republic, Denmark, Germany, the Netherlands, Norway and Poland), the Civil-Military Cooperation (CIMIC) Group North, formally received its status as an international military headquarters of NATO on 15 January 2003.

So, its role is to provide both NATO commanders and civilian institutions with the necessary expertise to create the conditions necessary to bridge the civil-military gap in military operations, particularly in the post-conflict peace-support operations when cooperation between all those involved is an essential factor in helping local authorities to rebuild social structures and restore normal living conditions for the local population. Civil-military cooperation may also be an important aspect of humanitarian, disaster relief or other civil emergency operations undertaken by national or international military forces

Located in Budel, the Netherlands, CIMIC Group North is functionally attached to Joint Force Command Headquarters Brunssum and is establishing working and training relationships with this headquarters. However, its role is to offer support to all NATO operations. Its main priority has been to establish an effective operational capability, focusing initially on training and education for CIMIC personnel. A similar initiative has been developed by Italy, Hungary and Greece to create a CIMIC Group South that will be attached to Joint Force Command Headquarters Naples.

H) Reserve forces:

Let us emphasise that the importance of reserve forces is growing with the multiplication of military operations conducted by the international community in the form, for instance, of "coalitions of the willing" and UN-mandated operations led by different international organisations.

The National Reserve Forces Committee (NRFC) is a central forum of the Alliance for reservist issues and has the task of preparing conceptual proposals and developing approaches for the Military Committee (MC) and member countries in this area.

We can affirm that the NRFC constantly takes up current problems with regard to preventive security and sets forth the interrelations with reservist issues as comprehensively and with as much foresight as possible so as to fulfil its role as an advisory body. It also serves as a forum for the exchange of information between individual NATO countries that deal with reservist matters and for the harmonisation of reserve forces, whenever possible, in accordance with the best practice principle. Since 1996, the NRFC has focused on strengthening the operational readiness of NATO reserve forces by broadening the exchange of information and employing reserve forces jointly with active forces. A number of key areas are being examined such as mobilisation systems, the requirements for training and follow-on training of reservists, and the motivation of reservists.

The NRFC was established in 1981. At present, almost all NATO countries are members of the committee, and the International Military Staff, Allied Command Operations and Allied Command Transformation are represented in it by liaison officers (and Australia by a permanent observer). It holds plenary conferences at least twice a year.

Besides, NRFC delegations are appointed by the respective national ministries of defence, and the national heads of delegations (HoD) are mostly heads of reserve or commissioners of reserve of Allied forces.

The NRFC provides guidance to the Confédération interalliée des officiers de réserve (Interallied Confederation of Reserve Officers, or CIOR), which brings together all existing reserve officer associations in NATO countries. The CIOR is a non-political, non-governmental, non-profit-making organisation dedicated to cooperation between the national reserve officers associations of NATO countries and to solidarity within the Atlantic Alliance.

Thus, the members of these associations are active as civilians in business, industrial, academic, political and other fields of professional life, in addition to their role as reserve officers. They are therefore in a position to contribute to a better understanding of security and defence issues in the population as a whole, as well as bringing civilian expertise and experience to the tasks and challenges facing reserve forces in NATO.

We want you to remember that the CIOR was founded in 1948 by the reserve officer associations of Belgium, France and the Netherlands. Its principal objectives include working to support the policies of NATO and to assist in the achievement of the Alliance's objectives, maintaining contacts with NATO's military authorities and commands, and developing international contacts between reserve officers in order to improve mutual knowledge and understanding.

Delegates to the CIOR are elected by their national reserve officer associations. The head of each delegation is a CIOR vice-president. The CIOR International President and Secretary General are elected by an Executive Committee, the CIOR's policy body that decides which country will assume the presidency, where congresses will be held, what projects will be assumed by the various commissions and the final actions to be taken on these projects.

The CIOR meets on an annual basis in the summer, alternating the location among member countries. It also organises a winter conference for the CIOR Executive Committee and Commissions.

In addition, the Confédération interalliée des officiers médicaux de réserve (Interallied Confederation of Medical Reserve Officers, or CIOMR) is an associated member of the CIOR. It holds its sessions at the same time and place as the CIOR summer congress and winter conference but follows its own agenda for the discussion of medical matters.

It is interesting to know that the CIOMR was established in 1947 as the official organisation of medical officers within reserve forces from countries which were to become NATO members. Originally founded by Belgium, France and the Netherlands, the Organization now includes all CIOR member countries. Its objectives include establishing close professional relations with the medical doctors and services of the reserve forces of NATO countries, studying issues of importance to medical reserve officers, including medico-military training, and promoting effective collaboration with the active forces of the Alliance.

Whenever possible the CIOR, the CIOMR and the NRFC convene at the same time and place. The three bodies also try to harmonise their respective programmes and projects.


**The international military staff's key functions**

In this section we will also follow the structure marked by the NATO texts and the philosophy that emerged from its complex institutional framework.

In the same way that the International Staff is the executive agency supporting the Council and its committees, so the International Military Staff (IMS), under the authority of its

Director, is the executive agency supporting the Military Committee.

The Director of the International Military Staff (DIMS) is a general or flag officer selected by the Military Committee from candidates nominated by member countries. Under his direction, the IMS prepares assessments, studies and reports that form the basis of discussion and decisions in the Military Committee. It is also responsible for planning, assessing and recommending policy on military matters for consideration by the Military Committee, and ensuring that the policies and decisions of the Committee are implemented as directed. The IMS provides the essential link between the political decision-making bodies of the Alliance and the Strategic Commanders and maintains close liaison with the civilian International Staff.

This way, the IMS consists of military personnel sent to take up staff appointments at NATO Headquarters, to work in an international capacity for the common interest of the Alliance rather than on behalf of their country of origin. Some posts within the IMS are filled by civilian personnel who work in administrative and support positions. As well as supporting the work of the Military Committee, preparing and following up its decisions, the IMS is also actively involved in the process of cooperation in the EAPC and PfP framework as well as the NATO-Russia Council, the NATO-Ukraine Commission and the Mediterranean Dialogue. Partner countries are represented within the International Military Staff and parts of the integrated command structure. These representations are called "Partnership for Peace Staff Elements". In times of tension, crisis and hostilities, or during NATO exercises, the IMS will implement a Crisis Management Organization based on functional military cells.

Coordination of staff action, and control of the flow of information and communications both within the IMS and between the IMS and other parts of the NATO Headquarters, is the responsibility of the Executive Coordinator, who works within the Office of the Director of the IMS.

Alliance regulations say that the Executive Coordinator and his/her staff also provide secretarial support and procedural advice to the Military Committee. A Public Information Adviser advises the Chairman of the Military Committee, the Deputy Chairman, and the Director of the IMS on public information matters, and acts as spokesperson for the Chairman and the Military Committee. This officer is also responsible for developing and monitoring military public information policy and doctrine. During NATO operations, the Public Information Adviser is the IMS representative in all committees and working groups dealing with public information matters and develops, coordinates and executes public information strategies. He/she works closely with the International Staff and the public information organisations within the Strategic Commands and the national ministries of defence.

Moreover, there is also a Financial Controller, who advises key officials on all financial and fiscal matters related to the group of budgets administered by the IMS. The Financial Controller is responsible to the Military Budget Committee for the financial management of the IMS budget. He/she is also tasked with preparing, justifying, administering and supervising all budget-related matters for presentation to the

Military Budget Committee, and for financial supervision of the NATO bodies with budgets administered by the IMS, namely the NATO Standardization Agency, the NATO Defence College, and the Research and Technology Agency. Finally, the financial controller is also responsible for conducting internal audits of accounts and activities with financial repercussions within his/her area of responsibility.

The Office of the Director of the IMS includes a Legal Counsel serving the Director and the IMS as a whole by providing advice on international and national legal implications of all aspects of NATO's military missions and of military advice provided by the Military Committee to the North Atlantic Council. Legal advice is given on legal aspects of operations, operations support, international laws and agreements relating to armed conflict, land, air and maritime operational plans, rules of engagement, targeting policy, the use of force, logistics and procurement matters, installations and other matters. The advice may also address NATO commitments with regard to non-NATO military and civilian entities as well as internal legal issues relating to the role of the IMS.

Finally, the Director of the International Military Staff is supported by assistant directors, each of whom is responsible for specific areas of activity.

A) Planning and policy:

One of the functions is to develop and coordinate the contribution of the Military Committee to NATO policy and planning matters, defence policy, strategic planning, special weapons policy planning, proliferation of weapons of mass destruction, defence and force planning, the NATO

Response Force (NRF) and the Prague Capabilities Commitments (PCC). Work is divided into three main areas dealing with strategic policy and concepts, nuclear, biological and chemical (NBC) policy, and defence and force planning. This includes contributing to the development of politico-military concepts, studies and assessments, NATO's defence and force planning process, the biannual defence review and long-term conceptual studies. The staff develops and represents the views of the Military Committee and of the NATO Strategic Commanders on military policy matters in different NATO bodies. They are also responsible for the conceptual development of documents related to NATO-EU relations and the NATO-EU strategic partnership, as well as the follow-up of Berlin Plus-related issues.

B) Regional cooperation and security:

As is obvious, cooperation and regional security is also a focus of the International Military Staff, involving military contacts and cooperation within the framework of the Euro-Atlantic Partnership Council and Partnership for Peace, the NATO-Russia Council, the NATO-Ukraine Commission and the Mediterranean Dialogue. Military advice on NATO involvement in different aspects of disarmament, arms control and cooperative security issues is also developed, as is cooperation with the Organization for Security and Co-operation in Europe (OSCE) in the field of disarmament, arms control and cooperative security. The International Military Staff also provides personnel for the Western Consultation Office (WCO) in Vienna, established to facilitate and enhance NATO's cooperation with the OSCE.

So, relations with Partner countries are integrated into the daily work of the IMS. Since 1994, a number of Partner country liaison offices and, since 1997, permanent diplomatic missions, have been opened at NATO Headquarters. Military links with Partner countries are further strengthened by Partnership for Peace Staff Elements, consisting of officers from NATO and Partner countries located within the IMS at NATO Headquarters as well as within the NATO integrated military structure. Officers from Partner countries filling such posts work alongside officers from NATO countries in an international capacity, participating in the preparation of policy discussions and the implementation of policy decisions dealing with relevant Partnership for Peace military matters.


C) Operations:

Equally, another important function of the International Military Staff is to support the Military Committee in the development of current operational plans and in addressing questions relating to the NATO force posture and other military management issues relating to NATO's role in international crises. This also includes the promotion and development of multinational training and exercises for NATO and PfP countries and coordinating efforts relating to the development of effective NATO information operations and the associated training and exercises. Support is also provided for the NATO Air Defence Committee and for air defence matters in general. A NATO liaison officer to the United Nations ensures regular contact with the UN on behalf of the International Military Staff and the Organization as a whole, when required.

D) Intelligence:

It is important to know that day-to-day strategic intelligence support is provided to the Secretary General, the North Atlantic Council and Defence Planning Committee, the Military Committee, and other NATO bodies. It is within the IMS that the collation and assessment of intelligence received from NATO member countries and NATO commands takes place and its dissemination within NATO Headquarters and to NATO commands, agencies, organisations and countries is centralised and coordinated.

Thus, NATO strategic intelligence estimates are produced and disseminated, intelligence policy documents and basic intelligence documents are managed and coordinated, and selected data bases and digital intelligence information services are maintained. Additional functions performed include strategic warning and crisis management roles, liaising with other NATO and national bodies performing specialised intelligence functions, informing NATO bodies of relevant developments and facilitating the formulation of military advice to NATO's political authorities.

E) Logistics and resources:

In this aspect, a number of important functions are fulfilled in the area of logistics and resources: the management of Alliance resources in support of NATO military bodies, the development and updating of military policy and procedures for the management of Alliance resources, and staff support to and appropriate representation of the Military Committee on the following:

- all matters concerning the development and assessment of NATO military policy and procedures for armaments, research and technology and Military Committee-related standardisation activities;

- all matters concerning logistics, medical, civil emergency planning, the military and civilian manpower and personnel function, and NATO common-funded resources provided by the NATO Security Investment Programme and the Military Budget;

- in conjunction with staff working on regional cooperation and security, all matters concerning logistics, armaments, including research and technology, resource management and Military Committee-related standardisation activities with all the countries and organisations involved in cooperation with NATO.

F)  Consultation, command and control (C3):

The NATO HQ Consultation, Command and Control Staff (NHQC3S) is a combined staff comprising civilian members of the International Staff and officers of the International Military Staff, serving the consultation, command and control requirements of the North Atlantic Council, the Military Committee and the NATO C3 Board (NC3B).

It is administratively located within the International Military Staff structure and works under the Director, NHQC3S, who is a Vice-Chairman of the NATO C3 Board and the Military Committee representative to the Board. Members of the NHQC3S provide support for the NC3B and its sub-committees and give advice to the Military Committee on C3/communication and information system

(CIS) policy standards, products, analysis and capability packages.

G) The NATO Situation Centre:

The NATO Situation Centre (SITCEN) assists the North Atlantic Council, the Defence Planning Committee and the Military Committee in fulfilling their respective functions in the field of consultation. The SITCEN serves as the focal point within the Alliance for the receipt, exchange and dissemination of political, military and economic information, and monitors political, military and economic matters of interest to NATO and to NATO member countries on a 24-hour basis. The SITCEN is also responsible for all NATO Headquarters external communications, both secure and non-secure, ensuring contact with national capitals, Strategic Commands and other international organisations, and for providing geographic support services for the NATO Headquarters. It also provides facilities to enable the rapid expansion of consultation during periods of tension and crises, and maintains and updates relevant background information during such periods.

H) The Office of the Women in NATO Forces:

This office provides the secretariat for, and acts as the adviser to, the Committee on Women in the NATO Forces (CWINF). It is responsible for developing a network with defence and other international agencies concerned with the employment of military women, providing briefings on gender integration, and collecting and managing relevant information for dissemination among member and Partner countries, Mediterranean Dialogue countries and other international agencies and researchers. The office serves as NATO's focal point for all issues relating to the recruitment and employment, training and development, and quality of life of women in uniform. It also examines their impact on levels of readiness and ability to work in a multinational environment during NATO-led missions.

# NATO SPECIALISED ORGANISATIONS, AGENCIES, COMMITTEES AND POLICY BODIES

## Specialised organisations and agencies

Besides to its political headquarters and military command structures, NATO also has a number of specialised agencies located in different NATO member countries.

Whereas the International Staff and International Military Staff cover the day-to-day activities of the Alliance as well as activities related to its political and military agenda, the agencies have responsibilities in more technical fields and areas of specialisation that complement and form an integral part of NATO's agenda.

They provide advice and undertake research, support the implementation of Alliance decisions, provide communications and information systems services, and manage cooperative programmes.

One of their main roles is to facilitate the best use of the limited defence resources of member countries through the development of common projects, procedures and standards.

Thus, at the Prague Summit meeting in November 2002, the NATO Secretary General was commissioned by NATO heads of state and government to undertake a review of the roles and requirements for NATO's agencies, including their relationship with NATO structures as a whole and with the North Atlantic Council in particular.

The aims of the review are to examine the effectiveness and coherence of the agency structure, opportunities for project rationalisation and improvements in reporting channels and coordination mechanisms, as well as the relationship between the agencies' roles and NATO's ongoing transformation process as a whole, including measures to bring about improvements in capabilities.

The review is being undertaken in parallel to the redirection of resources and rationalisation of the military structure agreed upon at the Prague Summit meeting, and in particular those aspects that impact upon the role of Atlantic Command Transformation, which has the leading role in relation to the development of military structures, forces, capabilities and doctrines, research and acquisition processes, interoperability and standardisation issues and training and education programmes.

To specify we will say that there are essentially two types of agencies, namely those that act as project coordinators and those that are service providers.

Several of the agencies are concerned with identifying the member countries' collective requirements and managing the production and logistics of common procurement projects on their behalf.

At one end of the scale are agencies managing major projects and therefore dealing with large budgets, such as the NATO Airborne Early Warning and Control Programme Management Agency (NAPMA), the NATO Air Command and Control System Management Agency (NACMA), the NATO Battlefield Information Collection and Exploitation System (BICES) Agency and the Central Europe Pipeline Management Agency (CEPMA).

Other logistics agencies are concerned with practical cooperation in all aspects of logistic support, including the purchase of logistic items and the maintenance of defence equipment. The main logistics agency in this field is the NATO Maintenance and Supply Agency (NAMSA) based in Luxembourg. It provides cost-effective logistic support for NATO weapon systems operated by 25 of NATO's member countries and helps NATO and its Partner countries to purchase items of equipment as well as spare parts and maintenance and repair facilities at the lowest possible cost.

It is necessary to know that the agencies report through Boards of Directors or Steering Committees to the North Atlantic Council under whose authority they normally operate.

Heads of agencies meet on a regular basis, hold meetings with the NATO Secretary General and receive briefings on the latest developments and thinking in defence procurement, planning and operations, personnel policy and security.

They are supported by their own staffs and coordinate their efforts in order to contribute to the overall process of moving forward on issues on NATO's current agenda, for example through developing standardisation practices and common codification systems.

In addition, the programmes and activities undertaken by the agencies vary considerably and require different forms of budgeting and financial cost-sharing arrangements reflecting their specific roles and varying membership.

Each agency is governed by its own specific charter, and its relationship with the country in which it is located and with the other participating countries is subject to specific memoranda of understanding.

We can also mention that the issues of standardisation and interoperability of forces from NATO and Partner countries remain high on NATO's agenda in view of its key role in facilitating multinational military operations. The NATO Standardization Agency works toward the implementation of common standards and the adaptation of procedures and practices necessary to achieve them.

It is interesting to know that the NATO Consultation, Command and Control Agency (NC3A), based in Brussels and The Hague, is another agency which has major responsibilities on behalf of the Alliance for the development of Allied capabilities in communications and information systems. Its role is to ensure that the command and control structures and forces of NATO and Partner countries are able to communicate together, especially during crises.

The agency deals with matters such as operational research, intelligence, surveillance, reconnaissance, air command and control, and communications and information systems. It provides central planning, systems architecture, systems integration, design, systems engineering, technical support and configuration control.

Moreover, collaborative studies in the field of scientific research are supported by the NATO Research and Technology Agency (RTA), based in Neuilly, France, on behalf of NATO's Military Committee and the Conference of National Armaments Directors (CNAD), which is responsible for cooperation in matters relating to defence acquisition.

There are also a number of specialised NATO agencies engaged in managing procurement programmes such as the NATO Medium Extended Air Defence System Design and Development, Production and Logistics Management Agency (NAMEADSMA), the NATO EF 2000 and Tornado Development Production and Logistics Management Agency (NETMA), the NATO Helicopter Design and Development, Production and Logistics Management Agency (NAHEMA) and the NATO HAWK Management Office (NHMO).

Other agencies and organisations are active in fields such as civil emergency planning, air traffic management and air defence, electronic warfare, meteorology and military oceanography.

In addition, there are a number of multinational institutions that play a key role in the field of education and training, such as the NATO Defence College in Rome, Italy, the NATO School in Oberammergau, Germany, and the NATO Communications and Information Systems School in Latina, Italy.

Regardless that this information may vary over time, more information can be found at the headquarters of these agencies:

NATO Battlefield Information Collection and Exploitation System (BICES) Agency Z Building – Blvd Léopold III 1110 Brussels, Belgium.

Central Europe Pipeline Management Agency (CEPMA) 11bis rue du Général Pershing, BP 552 78005 Versailles CEDEX, France.
Common Regional Initial ACCS Programme/Regional Programme Office (CRIAP/RPO) Quartier Reine Elisabeth Bloc 5A Rue d'Evere 1140 Brussels, Belgium.

NATO ACCS Management Agency (NACMA) Z Building – Blvd Léopold III 1110 Brussels, Belgium.

NATO Helicopter Design and Development, Production and Logistics Management Agency (NAHEMA) Le Quatuor Bâtiment A-42 Route de Galice 13090 Aix-en-Provence, France.

NATO Medium Extended Air Defence System Design and Development, Production and Logistics Management Agency (NAMEADSMA) 620 Discovery Drive Building 1 - Suite 300 Huntsville, AL 35806, USA.

NATO Maintenance and Supply Agency (NAMSA) 8302 Capellen, Luxembourg.

NATO Consultation, Command and Control Agency (NC3A) NC3A Brussels, Z Building Blvd Léopold III, 1110 Brussels, Belgium. And NC3A The Hague, PO Box 174, 2501 CD The Hague, Netherlands.

NATO Airborne Early Warning and Control Programme Management Agency (NAPMA) Akerstraat, 6445 CL Brunssum, Netherlands.

NATO Defence College (NDC) Via Giorgio Pelosi 1 00143 Rome, Italy.

NATO Standardization Agency (NSA) NATO Headquarters 1110 Brussels, Belgium.

Research and Technology Agency (RTA) BP 25, F-92201 Neuilly-sur-Seine CEDEX, France.

NATO CIS Services Agency (NCSA) SHAPE, B-7010, Belgium.


**Key to the principal NATO committees**

Let us start by saying that the principal forums for Alliance consultation and decision-making are supported by a committee structure which ensures that each member country is represented at every level in all fields of NATO activity in which it participates.

Some of the committees were established in the early days of NATO's development and have contributed to the Alliance's decision-making process for many years. Others have been established more recently in the context of the Alliance's internal and external adaptation, following the end of the Cold War and the changed security environment in Europe.

The Secretary General is the titular chairman of a number of policy committees which are chaired or co-chaired on a permanent basis by senior officials responsible for the subject area concerned.

It must be noted that the denomination of the divisions for which certain Assistant Secretary Generals or Deputy Assistant Secretary Generals are responsible can change following reforms of the International Staff, which take place on a regular basis.

The main source of support shown under the respective committees is the division of the International Staff with the primary responsibility for the subject matter concerned. Many of the committees are also supported by the International Military Staff.

All NATO committees take decisions or formulate recommendations to higher authorities on the basis of exchanges of information and consultations leading to consensus. There is no voting or decision by majority.

We must review that the NATO Military Committee is subordinate to the North Atlantic Council and Defence Planning Committee but has a special status as the senior military authority in NATO.

Finally, it is interesting to know that the Military Committee and most of the Committees listed below also meet regularly with representatives of Partner countries in the framework of the Euro-Atlantic Partnership Council (EAPC) and with representatives of Mediterranean Dialogue countries. In this context, the specific committees to which we refer are the following:

1. North Atlantic Council (NAC)

2. Defence Planning Committee (DPC)

3. Nuclear Planning Group (NPG)

4. Military Committee (MC)

5. Executive Working Group (EWG)

6. High Level Task Force on Conventional Arms Control (HLTF)

7. Joint Committee on Proliferation (JCP)

8. Political-Military Steering Committee on Partnership for Peace (PfP/SC)

9. NATO Air Defence Committee (NADC)

10. NATO Consultation, Command and Control (C3) Board (NC3B)

11. NATO Air Command and Control System (ACCS) Management Organization Board of Directors (NACMO BoD)

12. Senior Political Committee (SPC)

13. Atlantic Policy Advisory Group (APAG)

14. Political Committee (PC)

15. Senior Politico-Military Group on Proliferation (SGP)

16. Verification Coordinating Committee (VCC)

17. Policy Coordination Group (PCG)

18. Defence Review Committee (DRC)

19. Conference of National Armaments Directors (CNAD)

20. NATO Committee for Standardization (NCS)

21. Infrastructure Committee

22. Senior Civil Emergency Planning Committee (SCEPC)

23. Senior NATO Logisticians' Conference (SNLC)

24. Science Committee (SCOM)

25. Committee on the Challenges of Modern Society (CCMS)

26. Civil and Military Budget Committees (CBC/MBC)

27. Senior Resource Board (SRB)

28. Senior Defence Group on Proliferation (DGP)

29. High Level Group (NPG/HLG)

30. Economic Committee (EC)

31. Committee on Public Diplomacy (CPD)

32. Council Operations and Exercises Committee (COEC)

33. NATO Air Traffic Management Committee (NATMC)

34. Central Europe Pipeline Management Organization Board of Directors (CEPMO BoD)

35. NATO Pipeline Committee (NPC)

36. NATO Security Committee (NSC)

37. Special Committee

38. Archives Committee

## Key to the institutions of cooperation, partnership and dialogue

Finally, the institutions of cooperation, partnership and dialogue that underpin relations between NATO and other countries are:

Euro-Atlantic Partnership Council (EAPC)

NATO-Russia Council (NRC)

NATO-Ukraine Commission (NUC)

Mediterranean Cooperation Group (MCG)

Istanbul Cooperation Initiative Group (ICIG)

# THE ALLIANCE'S ROLE IN PEACEKEEPING AND PEACE-SUPPORT OPERATIONS

## NATO's role in Bosnia and Herzegovina

After the end of the Cold War, NATO has become increasingly involved in peacekeeping and peace-support operations, deploying in support of the wider interests of the international community and working closely together with other organisations to help resolve deep-rooted problems, alleviate suffering and create the conditions in which peace processes can become self-sustaining.

NATO's first three peace-support operations took place in Europe (in Bosnia and Herzegovina, in Kosovo and in the former Yugoslav Republic of Macedonia) yet the need for long-term peace-building is global. NATO foreign ministers recognised this at a meeting in Reykjavik, Iceland, in May 2002 agreeing that: "To carry out the full range of its missions, NATO must be able to field forces that can move quickly to wherever they are needed, sustain operations over distance and time, and achieve their objectives." This decision effectively paved the way for NATO to deploy for the first time outside the Euro-Atlantic area, in Afghanistan in 2003. Since then, the Alliance became involved in both Iraq and in Darfur, Sudan.

Bosnia and Herzegovina has been important to NATO because this territory was the scene of many firsts for NATO, and decisions taken in response to events in that country have helped shape the Alliance's evolution and develop its peacekeeping and peace-support capabilities.

The Alliance carried out an air campaign in Bosnia and Herzegovina in August and September 1995 that helped bring the Bosnian War to an end and then led a peacekeeping operation there for nine years, from December 1995 to December 2004. Although NATO handed responsibility for ensuring day-today security in Bosnia and Herzegovina to the European Union in December 2004, the Alliance retains a residual military headquarters in Sarajevo to focus on defence reform in Bosnia and Herzegovina and prepare the country for membership of the Partnership for Peace programme.

it is important to know that the political basis for the Alliance's role in peacekeeping operations was established at an Oslo meeting of NATO foreign ministers in June 1992. At that meeting, the foreign ministers announced their readiness to support peacekeeping activities under the responsibility of the Conference on Security and Co-operation in Europe (CSCE, subsequently renamed the Organization for Security and Co-operation in Europe, or OSCE) on a case-by-case basis and in accordance with their own procedures. This included making Alliance resources and expertise available for peacekeeping operations.

Then, in December 1992, the Alliance stated that it was also ready to support peacekeeping operations under the authority of the UN Security Council, which has primary responsibility for international peace and security. Reviewing the peacekeeping and sanctions or embargo enforcement measures already being undertaken by NATO countries, individually and as an Alliance, to support the implementation of UN Security Council resolutions relating to the conflict in the former Yugoslavia, NATO foreign ministers indicated that the Alliance was ready to respond positively to further initiatives that the UN Secretary-General might take in seeking Alliance assistance in this field.

And between 1992 and 1995, the Alliance took several key decisions which led to operations to monitor, and subsequently enforce, a UN embargo and sanctions in the Adriatic and to monitor and then to enforce the UN no-fly zone over Bosnia and Herzegovina. The Alliance also provided close air support to the UN Protection Force (UNPROFOR) and authorised air strikes to relieve the siege of Sarajevo and other threatened areas designated by the United Nations as safe areas.

On 30 August 1995, NATO aircraft launched a series of precision strikes against selected targets in Serb-held positions in Bosnia and Herzegovina. This heralded the start of Operation Deliberate Force, NATO's first air campaign, which lasted until 15 September. The operation shattered Bosnian Serb communications and, in conjunction with a determined diplomatic effort, helped pave the way to a genuine cease-fire; moreover, it prepared the ground for successful peace negotiations in Dayton, Ohio, United States.

Dayton Peace Accord:

Under the terms of the General Framework Agreement for Peace in Bosnia and Herzegovina, commonly referred to as the Dayton Peace Accord (DPA), signed in Paris on 14 December 1995, a NATO-led Implementation Force (IFOR) of 60 000 troops was established for one year to oversee implementation of the military aspects of the agreement. The Force was activated on 16 December, and transfer of authority from the Commander of UN forces to the Commander of IFOR took place four days later, bringing all NATO and non-NATO forces participating in the operation under IFOR command.

Finally, by 19 January 1996, the parties to the DPA had withdrawn their forces from the zone of separation on either side of the agreed cease-fire line and by 3 February, all forces had been withdrawn from the areas to be transferred under the terms of the Agreement. The transfer of territory between the entities of Bosnia and Herzegovina was completed by 19 March and a new zone of separation was established. By the end of June, the cantonment of heavy weapons and demobilisation of forces required under the DPA had also been completed. After more than four years of conflict and the repeated failure of international initiatives to end it, a basis for the future peace and security of Bosnia and Herzegovina had been established within less than six months.

This way, IFOR contributed substantially to the creation of a secure environment conducive to civil and political reconstruction. It also provided support for civilian tasks, working closely with the Office of the High Representative (OHR), the International Police Task Force (IPTF), the International Committee of the Red Cross (ICRC), the office of the UN High Commissioner for Refugees (UNHCR), the International Criminal Tribunal for the former Yugoslavia (ICTY) and many other agencies, including more than 400 non-governmental organisations active in the area.

IFOR also assisted the Organization for Security and Co-operation in Europe (OSCE) in preparing, supervising and monitoring the first free elections in September 1996 and, following those elections, supported the OHR in assisting the entities of Bosnia and Herzegovina in building new common institutions. In addition, IFOR military engineers repaired and reopened roads and bridges and played a vital role in demining efforts, repairing railroads, opening up airports to civilian traffic, restoring gas, water and electricity supplies, rebuilding schools and hospitals, and restoring key telecommunication installations.

From IFOR to SFOR:

In November and December 1996, a two-year consolidation plan was established under the auspices of the Peace Implementation Council, an ad hoc group consisting of countries and international organisations with a stake in the peace process. On the basis of this plan and of the Alliance's own study of security options, NATO foreign and defence ministers concluded that a reduced military presence was needed to provide the stability necessary for consolidating peace in the area. They agreed that NATO should organise and lead a 32 000-strong Stabilisation Force (SFOR), which was subsequently activated on 20 December 1996 – the day on which IFOR's mandate expired – with a new 18-month mandate.

Thus, in accordance with UN Security Council Resolution 1088 of 12 December 1996, SFOR became the legal successor to IFOR, its primary task being to contribute to the development of the secure environment necessary for the consolidation of peace. A further follow-on force retained the name "SFOR" and continued to operate on a similar basis, in order to deter renewed hostilities and to help create the conditions needed for the implementation of the civil aspects of the DPA. At the same time, the North Atlantic

Council projected a transitional strategy involving progressive reductions of force levels as the transfer of responsibilities to the competent common institutions, civil authorities and international bodies became feasible.

Over time, as the situation in Bosnia and Herzegovina became more stable, NATO restructured and reduced the size of the Stabilisation Force. By the beginning of 2002, it had been reduced from its original 32 000 troops to approximately 19 000 drawn from 17 NATO member countries and 15 non-NATO countries, including a Russian contingent. A large number of non-NATO countries, some of which have since become members, participated in IFOR and SFOR at different times, including Albania, Argentina, Austria, Bulgaria, Egypt, Estonia, Finland, Ireland, Jordan, Latvia, Lithuania, Morocco, Romania, Slovakia, Slovenia, Sweden and Ukraine.

SFOR was further reduced to 12 000 troops by January 2003, with the support of strategic reserve forces if required and a continuing mandate to help maintain a safe and secure environment in accordance with the DPA. Improvements in the overall security situation in Bosnia and Herzegovina in 2003, including successful operations conducted by explosive ordnance disposal units to destroy large quantities of grenades, rifles, pistols, mines and other munitions, enabled NATO further to reduce SFOR's size to a residual deterrent force of some 7000 troops, once again backed by reinforcement possibilities, by mid-2004.

At the same time, the successful handover to the European Union of the NATO operation in the former Yugoslav Republic of Macedonia in 2003 opened the way for the deployment of an EU follow-on mission to succeed SFOR.

Recognising the progress made in Bosnia and Herzegovina since the deployment of the NATO-led Implementation Force in 1995 as well as the subsequent positive role undertaken by the SFOR, Alliance leaders agreed to conclude the SFOR operation by the end of 2004.

One more step was when, on 2 December 2004, the European Union deployed a new force in Bosnia and Herzegovina, EUFOR, in Operation Althea. EUFOR benefits from ongoing NATO support in accordance with the Berlin-Plus arrangements made between the two organisations. Preparations for the transfer of responsibility for this mission were undertaken in the framework of these arrangements, drawing on NATO planning expertise and paving the way for the use by the European Union of the Alliance's collective assets and capabilities. In particular, the provisions enabled the Deputy Supreme Allied Commander Europe (DSACEUR) to become the Operation Althea Commander. These arrangements also enabled the transition of responsibility for the mission from NATO to the European Union to take place without interruption, which optimised the use of resources and avoided duplicating efforts.

Even if NATO's role as the main provider of security in Bosnia and Herzegovina concluded with the completion of the SFOR mission, the Alliance's continuing commitment to the country manifests itself in other ways. On 2 December 2004, the Alliance established a military headquarters in the country as a residual military presence to help the national authorities as they tackle the problems of defence reform and prepare for possible future participation in the Partnership for Peace programme. The headquarters has also undertaken certain operational support tasks such as counter-terrorism; supporting the International Criminal Tribunal for the former Yugoslavia (ICTY), within the means and capabilities at the headquarters' disposal, with the detention of persons indicted for war crimes; and intelligence-sharing with the European Union.

NATO continued to demonstrate its practical support for Bosnia and Herzegovina's efforts to join the Partnership for Peace (PfP) and the Euro-Atlantic Partnership Council through activities organised in the framework of a concrete NATO Security Cooperation Programme with that country.

To summarise, we can specify the operations to support peace in Bosnia and Herzegovina carried out by the Alliance at the following points:

- ✓ NATO conducted its first major crisis-response operation in Bosnia and Herzegovina.

- ✓ NATO implemented the military aspects of the Dayton Peace Agreement, which marked the end of the 1992-1995 war in the country.

- ✓ The NATO-led Implementation Force (IFOR) was deployed in December 1995 and was followed by the NATO-led Stabilization Force (SFOR), which ended in December 2004.

- ✓ Once NATO had successfully implemented the military aspects of the Dayton Peace Agreement, the European Union (EU) took on NATO's stabilisation role.

✓ NATO maintains a military headquarters in Sarajevo that complements the work of the EU mission and assists, inter alia, in defence reform and counter-terrorism.

✓ Bosnia and Herzegovina became a NATO partner country in December 2006 and is focusing on introducing democratic, institutional and defence reforms, as well as developing practical cooperation in other areas.

**The Kosovo conflict and the role of KFOR**

Let us start by saying that NATO has been leading a peacekeeping operation in Kosovo since June 1999 in support of wider international efforts to build peace and stability in the contested province. The NATO-led Kosovo Force, or KFOR, deployed in the wake of a 78-day air campaign launched by the Alliance in March 1999 to halt and reverse the humanitarian catastrophe that was then unfolding. That campaign, which was NATO's second, followed more than a year of fighting in the province and the failure of international efforts to resolve the conflict by diplomatic means.

Unfortunately, simmering tension in Kosovo resulting from the 1989 imposition of direct rule from Belgrade of this predominantly Albanian province erupted in violence between Serbian military and police and Kosovar Albanians at the end of February 1998.

The international community became increasingly concerned about the escalating conflict, its humanitarian consequences and the risk of it spreading to other countries, as well as Yugoslav President Slobodan Milosevic's disregard for diplomatic efforts aimed at peacefully resolving the crisis and the destabilising role of Kosovar Albanian militants.

Thus, on 13 October 1998, the North Atlantic Council authorised activation orders for NATO air strikes, in support of diplomatic efforts to make the Milosevic regime withdraw forces from Kosovo, cooperate in bringing an end to the violence and facilitate the return of refugees to their homes. Following further diplomatic initiatives, President Milosevic agreed to comply and the air strikes were called off.

Further measures were taken in support of UN Security Council resolutions calling for an end to the conflict, including the establishment of a Kosovo Verification Mission by the Organization for Security and Co-operation in Europe (OSCE) and an aerial surveillance mission by NATO, as well as a NATO military task force to assist in the evacuation of members of the Verification Mission in the event of further conflict.

At the beginning of 1999, the situation in Kosovo flared up again, following a number of acts of provocation on both sides and the use of excessive force by the Serbian military and police. This included the massacre of 40 unarmed civilians in the village of Racak on 15 January. Renewed international efforts to give new political impetus to finding a peaceful solution to the conflict resulted in the convening of negotiations between the parties to the conflict in London and Paris under international mediation.

These negotiations failed, however, and in March 1999, Serbian military and police forces stepped up the intensity of their operations, moving extra troops and tanks into the region, in a clear breach of agreements reached. Tens of thousands of people began to flee their homes in the face of this systematic offensive. A final unsuccessful attempt was made by US Ambassador Richard Holbrooke to persuade President Milosevic to reverse his policies. All diplomatic avenues having been exhausted, NATO launched an air campaign against the Milosevic regime on 24 March 1999.

We should remember that NATO's political objectives were to bring about a verifiable stop to all military action, violence and repression; the withdrawal from Kosovo of military personnel, police and paramilitary forces; the stationing in Kosovo of an international military presence; the unconditional and safe return of all refugees and displaced persons and unhindered access to them by humanitarian aid organisations; and the establishment of a political agreement for Kosovo in conformity with international law and the Charter of the United Nations.

In the end, following diplomatic efforts by Russia and the European Union on 3 June, a Military Technical Agreement was concluded between NATO and the Federal Republic of Yugoslavia on 9 June. On the following day, after confirmation that the withdrawal of Yugoslav forces from Kosovo had begun, NATO announced the suspension of the air campaign. On 10 June, UN Security Council Resolution 1244 welcomed the Federal Republic of Yugoslavia's acceptance of the principles for a political solution, including an immediate end to violence and a rapid withdrawal of its military, police and paramilitary forces and the deployment of an effective international civil and security presence, with substantial NATO participation.

The NATO-led Kosovo Force:

On 12 June 1999, the first elements of KFOR entered Kosovo. By 20 June, the withdrawal of Serbian forces was complete. KFOR tasks have included assistance with the return or relocation of displaced persons and refugees; reconstruction and demining; medical assistance; security and public order; security of ethnic minorities; protection of patrimonial sites; border security; interdiction of cross-border weapons smuggling; implementation of a Kosovowide weapons, ammunition and explosives amnesty programme; weapons destruction; and support for the establishment of civilian institutions, law and order, the judicial and penal system, the electoral process and other aspects of the political, economic and social life of the province.

KFOR was initially composed of some 50 000 personnel from NATO member countries, Partner countries and non-NATO countries under unified command and control. By the beginning of 2002, KFOR had been reduced to around 39 000 troops. Improvements in the security environment enabled NATO to reduce KFOR troop levels to around 26 000 by June 2003 and to 17 500 by the end of that year. A setback in progress towards a stable, multi-ethnic and democratic Kosovo occurred in March 2004, when renewed violence broke out between Albanians and Serbs and KFOR troops were attacked. NATO contingency plans for such an eventuality enabled the rapid deployment of some 2500 additional troops to reinforce the existing KFOR strength.

At the Istanbul Summit, NATO heads of state and government condemned the renewed ethnic violence that had erupted in March 2004 and reaffirmed NATO's commitment to a secure, stable and multi-ethnic Kosovo, on the basis of full implementation of United Nations Security Council Resolution 1244. They also reiterated their support for the agreed "Standards before Status" policy and the associated Standards Review Mechanism.

Ahead of the comprehensive review of the Standards Implementation Process scheduled for the end of 2005, NATO defence ministers agreed at their meeting in Brussels in December 2004 to maintain a robust KFOR profile during the year 2005. In the meantime, in August 2005, the North Atlantic Council decided to restructure KFOR, replacing the four existing multinational brigades with five task forces. This reform will be introduced gradually and will allow greater flexibility with, for instance, the removal of restrictions on the cross-boundary movement of units based in different sectors of Kosovo. The move from brigade to task force will also place more emphasis on intelligence-led operations, with task forces working closely with both the local police and the local population to gather information.

Support for neighbouring countries:

Likewise, as a result of the conflict in Kosovo, the countries of the region faced major humanitarian, political and economic problems. At the height of the Kosovo crisis, more than 230 000 refugees had arrived in the former Yugoslav Republic of Macedonia,* more than 430 000 in Albania and some 64 000 in Montenegro. Approximately 21 500 had reached Bosnia and Herzegovina and more than 61 000 had been evacuated to other countries. Within Kosovo itself, an estimated 580 000 people had been rendered homeless. To help ease the humanitarian situation on the ground, NATO forces flew in many thousands of tons of food and equipment. By the end of May 1999, over 4666 tons of food and water, 4325 tons of other goods, 2624 tons of tents and nearly 1600 tons of medical supplies had been transported to the area.

Besides, in the former Yugoslav Republic of Macedonia, NATO troops built refugee camps, refugee reception centres and emergency feeding stations and moved hundreds of tons of humanitarian aid to those in need. In Albania, NATO deployed substantial forces to provide similar forms of assistance and helped the UNHCR with the coordination of humanitarian aid flights to enable the evacuation of refugees to safety in other countries, including many NATO countries. These flights were supplemented by aircraft supplied by NATO member countries. The Euro-Atlantic Disaster Response Coordination Centre (EADRCC) established at NATO in June 1998 also played an important role in the coordination of support to UNHCR relief operations.

Later, a NATO PfP Cell was set up in Tirana from 1998 to December 2002 to assist the government with PfP programmes and procedures. In June 2002, NATO nominated a Senior Military Representative to Albania, with headquarters in Tirana. The role of the Senior Military Representative is to advise Tirana on military aspects of security sector reform, including the restructuring of the Albanian armed forces, and on military aspects of the Membership Action Plan and PfP Planning and Review Process, in both of which Albania is a participant. NATO Headquarters Tirana includes a NATO Advisory Team which assists the Senior Military Representative in the implementation of these tasks. A further task assigned to NATO Headquarters Tirana has been to provide support for NATO-led operations in the region. A significant contribution to NATO operations is also made by Albania itself, through the authorisation of surveillance and reconnaissance flights over its territory as well as cooperation on border security issues between Albanian border police and military units and KFOR.

When Kosovo proclaimed its unilateral independence in 2008, the countries that recognised it and NATO itself as such already had a stabilisation mission to protect the Albanian majority territory. Once independence was proclaimed, they decided to create a Kosovo Security Force (KSF) of limited dimensions and, essentially, to give a way out to the guerrillas who had fought against Serbia. Alliance Secretary-General Jens Stoltenberg, when Kosovo decided to create his own army, regretted that the decision was made despite the concerns expressed by NATO which he clearly stated that he considers this movement as inopportune.

To summarise, we can specify the operations to support peace in Kosovo carried out by the Alliance at the following points:

- ✓ NATO has been leading a peace-support operation in Kosovo - the Kosovo Force (KFOR) - since June 1999.

- ✓ KFOR was established when NATO's 78-day air campaign against Milosevic's regime, aimed at putting an end to violence in Kosovo, was over.

✓ The operation derives its mandate from United Nations Security Council Resolution 1244 (1999) and the Military-Technical Agreement between NATO, the Federal Republic of Yugoslavia and Serbia.

✓ KFOR's original objectives were to deter renewed hostilities, establish a secure environment and ensure public safety and order, demilitarise the Kosovo Liberation Army, support the international humanitarian effort and coordinate with the international civil presence.

✓ Today, KFOR continues to contribute towards maintaining a safe and secure environment in Kosovo and freedom of movement for all.

✓ NATO strongly supports the Belgrade-Pristina EU-brokered Normalisation Agreement (2013).

**NATO's role in the former Yugoslav republic of Macedonia**

It is necessary to remember that NATO became involved in the former Yugoslav Republic of Macedonia (now officially the Republic of North Macedonia) at the request of the Skopje authorities to help defuse an escalating conflict between the government and ethnic Albanian rebels to head off what might have degenerated into a full-scale war.

This way, in June 2001, President Boris Trajkovski of the former Yugoslav Republic of Macedonia asked for NATO assistance to help demilitarise the National Liberation Army

(NLA) and disarm ethnic Albanian groups operating on the territory of his country.

In response, the North Atlantic Council took a double-track approach: it condemned the attacks and adopted measures in support of the government's action against extremist activities, while urging the government to moderate its military action and adopt constitutional reforms to increase the participation of ethnic Albanians in society and politics.

Thus, a political dialogue between both parties was engaged, leading to a peace plan and a cease-fire. The signing of the Ohrid Framework Agreement on 13 August 2001 opened the way for the entry of NATO troops into the country on 27 August 2001 and for the introduction of internal reforms. The 30-day mission, code-named Operation Essential Harvest, was to collect and destroy all weapons voluntarily handed in by NLA personnel.

The operation involved some 3500 NATO troops and their logistical support. Approximately 3875 weapons and 397 600 other items, including mines and explosives, were collected. Later in the year, the 15 constitutional amendments in the peace agreement were passed by the Parliament.

And in September 2001, President Trajkovski requested a follow-on force to provide protection for international monitors from the European Union and the Organization for Security and Co-operation in Europe overseeing implementation of the peace plan for the former Yugoslav Republic of Macedonia. Known as Operation Amber Fox, the follow-on mission involved some 700 troops provided by NATO member countries, under German leadership, reinforcing some 300 troops already based in the country. It started on 27 September 2001 with a three-month mandate to contribute to the protection of international monitors overseeing the implementation of the peace plan and was subsequently extended.

So, in response to a request from President Trajkovski, NATO agreed to continue supporting the former Yugoslav Republic of Macedonia with a new mission starting on 16 December 2002, known as Operation Allied Harmony.

The North Atlantic Council recognised that while Operation Amber Fox could now be concluded, a follow-on international military presence in the country was still required to minimise the risk of destabilisation. The mission consisted of operational elements to provide support for the international monitors and advisory elements to assist the government in assuming responsibility for security throughout the country.

The NATO-led Operation Allied Harmony continued until 31 March 2003, when responsibility for the mission was handed to the European Union. NATO subsequently maintained both a civilian and a military presence in the country to assist and advise the national authorities on developing security sector reforms and on the country's participation in the Membership Action Plan (MAP).

NATO Headquarters Skopje, established for this purpose, consists of some 120 combined military and civilian personnel. It is a non-tactical headquarters under the command of a NATO Senior Military Representative. In the light of the damage and wear and tear on roads and bridges caused by increased military traffic and the use of the road network as military supply routes, NATO is also contributing to reconstruction and other civil engineering projects in the country.

NATO Headquarters Skopje plays an important role in the coordination of these efforts, which are being undertaken in conjunction with the civil engineering department of Skopje University.

The Prespa agreement, which replaces the Interim Accord of 1995, was signed on June 17, 2018 by the two Macedonian and Greek Foreign Ministers, Nikola Dimitrov and Nikos Kotzias, and in the presence of the respective Prime Ministers, Zoran Zaev and Alexis Tsipras, and the country's name was changed to Republic of North Macedonia.

In March 2020, after the ratification process by all NATO members was completed, North Macedonia acceded to NATO becoming the 30th member state.

The same month, the leaders of the European Union formally gave approval to North Macedonia begin talks to join the EU.

To summarise, we can specify the operations to support peace in North Macedonia carried out by the Alliance at the following points:

- ✓ With the signing of the Ohrid Framework Agreement (13 August 2001), the Skopje government pledged to improve the rights of its ethnic Albanian population and the latter agreed to abandon separatist demands and hand over weapons to a NATO force. This was the beginning of NATO's short-term military presence in the country (2001-2003).

✓ Operation Essential Harvest (22 August – 26 September 2001) helped to disarm ethnic Albanian extremists on a voluntary basis.

✓ Operation Amber Fox (27 September 2001 – 15 December 2002) was mandated to ensure the protection of international monitors from the European Union and the Organization for Security and Co-operation in Europe, which oversaw the implementation of the Ohrid Agreement.

✓ Operation Allied Harmony (16 December 2002 – 31 March 2003) provided continued support for the international monitors and assisted the government in taking ownership of security throughout the country.

✓ NATO maintained a military headquarters in Skopje to provide support in security sector reform until 2012, when the headquarters became the NATO Liaison Office Skopje.

✓ North Macedonia had been a NATO partner country since 1995 and, after a mutually acceptable solution to the issue of its name was reached with Greece, it became a NATO member in March 2020.

✓ The country has been recognised as the Republic of North Macedonia since 15 February 2019.

## NATO's role in Afghanistan

As is known, NATO leaded international peacekeeping efforts in Afghanistan since August 2003, thereby helping to establish the conditions in which the country can enjoy a representative government and self-sustaining peace and security. This groundbreaking operation is NATO's first beyond the Euro-Atlantic area. Initially restricted to providing security in and around Kabul, the Alliance expanded the mission to cover other parts of the country via so-called Provincial Reconstruction Teams. Specifically, NATO was seeking to assist the government of Afghanistan in maintaining security within its area of operations, to support the government in expanding its authority over the whole country, and to help provide a safe and secure environment conducive to free and fair elections, the spread of the rule of law, and the process of reconstruction.

Thus, in the wake of the ouster of al Qaida and the Taliban, Afghan leaders met in Bonn, Germany, in December 2001 with international backing to begin the process of rebuilding the country. A new government structure was created in the form of an Afghan Transitional Authority, and an International Security Assistance Force (ISAF) was created under United Nations Security Council Resolutions 1386, 1413 and 1444 to enable the Transitional Authority itself and the UN Assistance Mission in Afghanistan to operate in the area of the capital, Kabul, and its surroundings with reasonable security. A detailed Military Technical Agreement between the ISAF Commander and the Afghan Transitional Authority provided further guidance for ISAF operations.

ISAF was initially led by the United Kingdom and then by Turkey. Germany and the Netherlands jointly took over leadership of ISAF in February 2003 and in doing so requested NATO support. In August 2003, the Alliance itself took responsibility for ISAF in such a way that the problem of identifying new countries willing and able to take over the leadership of the mission every six months was overcome.

Besides, the international composition of ISAF has varied but, since its establishment, has included forces or contributions from all 26 NATO Allies and from Albania, Azerbaijan, Croatia, Estonia, Finland, Georgia, Ireland, New Zealand, Sweden, Switzerland and the former Yugoslav Republic of Macedonia, in addition to elements provided by Afghanistan itself.

ISAF's political direction was provided by the North Atlantic Council in consultation with non-NATO troop-contributing countries. NATO's Allied Command Operations (based at Supreme Headquarters Allied Powers in Europe located in Mons, Belgium), has responsibility for the operation's headquarters; Allied Joint Force Command Brunssum, in the Netherlands, acts as the operational-level headquarters.

In the first moment, the core of the ISAF headquarters in Kabul was formed from the Joint Command Centre in Heidelberg, Germany, which provided the first NATO ISAF Force Commander. Subsequently, command passed to Canada, then to the Eurocorps under French command, then to Turkey and then Italy. Together with its civilian support elements, the overall strength of ISAF amounts to approximately 8 000 personnel. A rotation plan has been developed that provides for the longer-term support of the ISAF's mission headquarters at least until February 2008.

In January 2004, NATO appointed former Turkish Foreign Minister Hikmet Cetin as its Senior Civilian Representative in Afghanistan, with responsibility for advancing political and military aspects of the Alliance's engagement in Afghanistan. The Senior Civilian Representative works under the guidance of the North Atlantic Council and in close co-ordination with the ISAF Commander and the UN Assistance Mission in Afghanistan, as well as with the Afghan authorities and other international bodies present in the country.

ISAF expansion:

UNSC Resolution 1510 opened the way for a wider role for ISAF to support the government of Afghanistan in regions of the country beyond the confines of the capital in October 2003. And in December 2003, the North Atlantic Council authorised NATO's Supreme Allied Commander Europe to initiate the expansion process.

Moreover, Provincial Reconstruction Teams, or PRTs, form the cornerstone of this process. They are teams composed of international civilian and military personnel structured as civil-military partnerships, the military elements of which are integrated into the ISAF chain of command. Their primary role is to help the government of Afghanistan extend its authority further afield and to facilitate the development of security in the regions. This includes establishing relationships with local authorities, enhancing security in their specific areas of operation, supporting security sector reform activities and using the means and capabilities available to them to help facilitate the reconstruction effort in the provinces.

The PRT concept is a new one which is proving to be an efficient and effective means of helping to create a secure environment and enabling lead countries, international organisations and non-governmental organisations to fulfil their own roles in assisting the government of Afghanistan to rebuild the country.

ISAF took over command of the German-led PRT in Kunduz as the pilot project and first step in the expansion process in December 2003. By the end of 2004, ISAF had taken command of the military components of five PRTs in the north of Afghanistan, located in Baghlan, Faizabad, Kunduz, Maymaneh and Mazar-e-Sharif. NATO also took responsibility for four PRTs in the west of the country, in Herat, Farah, Chagcharan and Qal'eh-Now, in mid-2005, bringing the total of NATO-led PRTs to nine, covering approximately 50 per cent of Afghanistan's territory. NATO has also decided to take over additional PRTs in the south and east of Afghanistan, which may necessitate greater synergy with the US-led Operation Enduring Freedom.

The composition and geographical reach of PRTs are determined by the NATO military authorities and the lead countries, in close consultation with the UN Assistance Mission to Afghanistan and the Afghan authorities and based on the specific situation in the provinces in which they operate. The specific objectives of individual PRTs take into account such factors as the local security situation, the status of reconstruction, and the presence of other international agencies.

Other components of ISAF:

- In addition to the PRTs, there are three other main components of ISAF. These are:

- The ISAF headquarters, which commands the Kabul Multinational Brigade and conducts operational tasks in its area of responsibility, liaising with and assisting in the work of the United Nations, the Afghan authorities, governmental and non-governmental organisations and the US-led coalition forces in Afghanistan (Operation Enduring Freedom).

- The Kabul Multinational Brigade, which is ISAF's tactical headquarters and is responsible for the planning and conduct of patrolling and civil-military cooperation operations on a day-to-day basis; and

- Kabul Afghan International Airport, which is operated by the Afghan Ministry of Civil Aviation and Tourism with the assistance of ISAF. NATO has an additional role in relation to the rehabilitation of Kabul airport, together with representatives of the other national and international bodies concerned.

ISAF also supported the conduct of the Constitutional Loya Jirga, or grand council, of some 500 Afghan leaders, which was held from December 2003 to early January 2004, and assisted the Afghan authorities in providing security for Kabul throughout the process. The ratification of the new constitution agreed by the Loya Jirga laid the foundation for the creation of democratic institutions and opened the way for free and fair national elections. In response to a request from Afghan President Hamid Karzai, ISAF also provided support during the presidential election period in autumn 2004 and the autumn 2005 parliamentary and local elections.

Let us say while primary responsibility for the conduct of the presidential elections rested with the Afghan government assisted by the UNAMA, additional forces were made available, including a Spanish Quick Reaction Force deployed to Marzar-e-Sharif and an Italian in-theatre reserve force located in Kabul.

Additional Dutch and UK aircraft and helicopter support was also provided, and a US battalion was on hand for rapid deployment to the area if required. Close coordination took place throughout with other national and international agencies on the spot, including the United Nations, the European Union and the Organization for Security and Co-operation in Europe.

The Bonn Agreement of December 2001 defines the institutional reforms required to lay the foundation for stability, peace and prosperity in five distinct spheres, namely counter narcotics; judicial reform; disarmament, demobilisation and reintegration; training of the Afghan National Army; and training of police forces. Lead donor countries from the G8 countries are assisting the Afghan authorities in carrying out security sector reform programmes in these spheres. Japan is the lead country overseeing the demobilisation, disarmament and reintegration process. The United States is leading international efforts to train the Afghan National Army. Germany has taken the lead in training the Afghan National Police. Italy is the lead country for judicial reform. The

United Kingdom is leading international efforts to help combat the production of and trade in narcotics.

Another aspect to point out is that within the framework of NATO-Russia cooperation, a joint pilot training project is also being developed to help build capacity in the region to more effectively tackle the trafficking in Afghan narcotics.

While the disarmament, demobilisation and reintegration process is not part of ISAF's mandate, its implementation impacts significantly on ISAF operations, particularly in and around Kabul. In March 2004, a ceremony outside Kabul marked the successful cantonment in safe storage sites of heavy weapons such as tanks, artillery pieces, surface-to-surface missiles and rocket-launching systems held by different militias in the capital. Initiated by the Afghan Ministry of Defence, the cantonment operates under a dual-key system and prevents the removal of these weapons without the agreement of both the Ministry and the ISAF Commander. A similar initiative implemented in the Panjsher Valley and the disarmament, demobilisation and reintegration process applied to armed groups in the country combine to form an integrated programme designed to bring the large number of weapons circulating in Afghanistan under control.

NATO is leading a non-combat mission to train, advise and assist the Afghan security forces and institutions. The Resolute Support Mission (RSM) was launched in January 2015, following the completion of the mission of the International Security Assistance Force (ISAF) in December 2014, when responsibility for security in Afghanistan was transferred to the Afghan national defence and security forces. Beyond supporting RSM, NATO Allies and partners are helping to sustain Afghan security forces and institutions financially, as part of a broader international commitment to Afghanistan. The NATO-Afghanistan Enduring Partnership provides a framework for wider political dialogue and practical cooperation.

To summarise, we can specify the operations to support peace in Afghanistan carried out by the Alliance at the following points:

- ✓ NATO led the UN-mandated International Security Assistance Force (ISAF) from August 2003 to December 2014. ISAF's mission was to enable the Afghan authorities and build the capacity of the Afghan national security forces to provide effective security, so as to ensure that Afghanistan would never again be a safe haven for terrorists.
- ✓ ISAF was NATO's longest and most challenging mission to date: at its height, the force was more than 130,000 strong with troops from 50 NATO and partner nations.

- ✓ ISAF also contributed to reconstruction and development in Afghanistan through 28 multinational Provincial Reconstruction Teams.

- ✓ The transition to Afghan lead for security started in 2011 and was completed in December 2014, when the ISAF operation ended and the Afghans assumed full responsibility for security of their country.

- ✓ In January 2015, NATO launched the Resolute Support Mission (RSM) to train, advise and assist Afghan security forces and institutions. Currently, it numbers around 17,000 troops from 39 NATO Allies and partner countries.

✓ At the July 2018 NATO Summit in Brussels, the Allies and their operational partners committed to sustain RSM until conditions indicate a change is appropriate; to extend financial sustainment of the Afghan security forces through 2024; and to make further progress on developing a political and practical partnership with Afghanistan.

✓ Set up in 2010, the Enduring Partnership is NATO's political partnership with Afghanistan. At the 2016 NATO Summit in Warsaw, Allies decided to strengthen and enhance the Partnership within and alongside RSM. In the longer term, a traditional partnership with Afghanistan remains NATO's goal.

✓ NATO's Senior Civilian Representative represents the political leadership of the Alliance in Kabul, advising the Afghan authorities on the Enduring Partnership as well as liaising with the government, civil society, representatives of the international community and neighbouring countries.

✓ NATO welcomes the announcement, on 29 February 2020, of significant first steps in pursuit of a peaceful settlement. Recent progress on peace has ushered in a reduction of violence and paved the way for intra-Afghan negotiations between a fully inclusive Afghan national team and the Taliban to reach a comprehensive peace agreement. The Allies call on the Taliban to embrace this opportunity for peace. In this context, conditions-based adjustments will be made to the Resolute Support Mission, including a reduced military presence.

## NATO's role in Iraq

NATO has become involved in various ways in helping with Iraq's transition since the end of the 2003 US-led campaign against Iraq and the ouster of the regime of Saddam Hussein. The Alliance is training Iraqi personnel both inside and outside Iraq and supporting the development of security institutions to help the country build effective armed forces and provide for its own security. NATO is also coordinating equipment donations to Iraq and providing support to Poland to help it command a sector in Iraq.

Thus, in May 2003, the North Atlantic Council agreed to provide Poland with assistance in the form of intelligence, logistics, movement coordination, force generation and secure communications. The decision was taken on a similar basis to the decision that had been taken to provide comparable forms of assistance to the Netherlands and Germany when they jointly assumed leadership of the International Security Assistance Force (ISAF) in Afghanistan. It came into effect immediately.

In June 2003, a Force Review Conference took place at Supreme Headquarters Allied Powers in Europe (SHAPE) with the participation of Poland, other NATO member countries and Partner countries to discuss force requirements and conclude arrangements for implementation. The conference forms part of the normal military planning process for any NATO operation and gives contributing countries the opportunity to discuss details, provide offers and finalise the force generation process. In September 2003, Poland assumed command of the Multinational Division (MND) Central South as part of the stabilisation force in Iraq. This role was reinforced by NATO as a whole as well as by bilateral contributions (including forces and other forms of support) by a number of individual NATO and Partner countries.

At the end of 2003 and at the beginning of 2004, statements issued on behalf of the North Atlantic Council emphasised that, without prejudice to subsequent decisions that might be taken in relation to the security situation in Iraq, the immediate operational priority for the Alliance remained the successful implementation of the role it had undertaken, from August 2003, in assuming command of the International Security Assistance Force in Afghanistan (ISAF). Ensuring effective implementation of this task would be a prerequisite for any subsequent decision relating to an enhanced Alliance role in relation to Iraq. However, the Alliance's role in relation to stabilisation efforts in Iraq would be kept under continuous review.

On 28 June 2004, sovereignty was formally transferred to an Interim Iraqi Government, the opening day of the NATO Istanbul Summit. In response to a request from the Iraqi Interim Government and following the unanimous adoption of UN Security Council Resolution 1546 asking international and regional organisations to assist the Multinational Force in Iraq, NATO leaders agreed to assist the Interim Government with the training of its security forces and tasked the North Atlantic Council to develop ways to implement this decision. Following discussions with the Interim Government, including visits to NATO by the Iraqi Foreign Minister in July and the Iraqi President in

September, it was also decided that NATO would provide further assistance with respect to the equipment and technical assistance for Iraq's security forces.

And on 30 July 2004, the Council agreed to the establishment of a NATO Training Implementation Mission numbering some 50 military personnel to begin training selected military and civilian headquarters personnel. Unlike operational missions involving combat forces, this was a distinct NATO training mission under the political control of the North Atlantic Council, working closely with the Iraqi authorities as well as with the US-led Multinational Force in Iraq. The aim of the mission is to help the Iraqi Interim Government to develop adequate national security structures as soon as possible, to provide for the future security of the Iraqi people. Security and protection for the mission itself is provided in part by the Multinational Force and in part by NATO.

This way, the specific tasks of the mission included establishing liaison arrangements with the Iraqi Interim Government and the Multinational Force; working with the Iraqi authorities to help them develop effective security structures, including training selected Iraqi headquarters personnel in Iraq; helping to identify Iraqi personnel for training outside Iraq; and working with the Interim Government and the Multinational Force to develop more detailed proposals for NATO training, advice and cooperation. Training and mentoring selected Iraqi personnel inside Iraq and developing a role in coordinating national offers of equipment and training began in August 2004.

The renamed NATO Training Mission is directed by an American general who is also in charge of the separate training programme led by the Multinational Force, thereby ensuring coordination while maintaining the distinct nature of the NATO programme. Overall responsibility for the programme rests with the Supreme Allied Commander, Operations, at SHAPE, who reports through the NATO Military Committee to the North Atlantic Council. SHAPE is supported by Allied Command Transformation in Norfolk, Virginia, United States, which is responsible for coordination of training efforts outside Iraq.

In September, based on the findings and recommendations of the NATO military authorities, NATO announced its intention to help create a NATO-supported Iraqi Training, Education and Doctrine Centre. Located near Baghdad, the role of the Centre is to focus on leadership training for Iraqi security forces and provide NATO assistance for the coordination of training being offered bilaterally by different member countries, both inside and outside Iraq.

The North Atlantic Council approved the Concept of Operations for the enhancement of NATO's assistance to the Iraqi Interim Government with the training of its security forces and with the coordination of offers of training and equipment in October 2004. The Concept of Operations provided the basis for a substantial practical enhancement of assistance within the framework of a distinct NATO mission and the development of a detailed Operations Plan which the North Atlantic Council approved in November 2004.

At the beginning of November 2004, 19 Iraqi security personnel participated in an eight-day training course at

NATO's Joint Warfare Centre at Stavanger, Norway – the first such training activity to be conducted outside Iraq and in accordance with the above decisions. The participants included senior military officers and civilian staff from the Iraqi Ministries of Defence and of the Interior. The course was designed to focus on the functioning of an operational-level headquarters and served as a pilot project for follow-on training both inside and outside Iraq.

Iraqi requests for further training by NATO or other organisations are coordinated by a NATO Training and Equipment Coordination Centre, which is working with a similar centre in Baghdad to coordinate the requirements of the Iraqi government for training and equipment with the support that is on offer by NATO as a whole and by individual NATO member countries.

When NATO foreign ministers met in Brussels in December 2004, they gave the formal go-ahead for the expansion of NATO's training assistance to Iraq. As a consequence, the NATO Training Mission was increased to some 300 training and support personnel, and the training and mentoring of senior Iraqi security personnel was stepped up.

From 2004 to 2011, NATO conducted a relatively small but important support operation in Iraq that consisted of training, mentoring and assisting the Iraqi security forces. It was known as the NATO Training Mission in Iraq (NTM-I) and became part of the international effort to help Iraq establish effective and accountable security forces. NTM-I delivered training, advice and mentoring support in a number of different settings. All NATO member countries contributed to the training effort either in or outside of Iraq, through financial contributions or donations of equipment.

In parallel and reinforcing this initiative, NATO also worked with the Iraqi government on a structured cooperation framework to develop the Alliance's long-term relationship with Iraq.

In July 2015, in response to a request by the Iraqi government, NATO agreed to provide defence and related security capacity building support. In April 2016, it began conducting a number of "train-the-trainer'' courses in Jordan (more than 350 Iraqi security and military personnel were trained). Then, following a request from the Iraqi Prime Minister, at the Warsaw Summit in July 2016, NATO leaders agreed to provide NATO training and capacity-building activities to Iraqi security and military forces in Iraq. In January 2017, NATO deployed a modest but scalable Core Team to Baghdad of eight civilian and military personnel, setting up NATO's permanent presence in Iraq. Jordan-based training transferred to Iraq in February 2017. The Core Team coordinated all NATO assistance provided to Iraq in 2017-2018 and laid the foundation for the establishment of NMI in 2018.

And to summarise, we can specify the latest operations to support peace in Iraq carried out by the Alliance at the following points:

- ✓ At the Brussels Summit in July 2018, NATO leaders agreed to launch NATO Mission Iraq (NMI), on the request of Iraq.

- ✓ NMI is a non-combat training and capacity-building mission, conducted with full respect of Iraq's sovereignty and territorial integrity.

✓ It was established in Baghdad in October 2018 and involves around 500 trainers, advisors and supporting personnel from Allied and partner countries, including Australia, Sweden and Finland.

**NATO's role in Africa: Cooperation with the African Union**

Together with the European Union, NATO has been assisting the African Union in expanding its peacekeeping mission in Darfur, Sudan, since July 2005 in an attempt to halt continuing violence. The Alliance has been airlifting African Union (AU) peacekeepers and civilian police into the war-ravaged region and providing training in running a multinational military headquarters and managing intelligence.

The African Union asked NATO in April 2005 to consider providing logistical support to help it expand its operation in Darfur, the African Union Mission in Sudan, to halt ongoing violence. In May, the Chairperson of the Commission of the African Union, Alpha Oumar Konaré, became the first African Union official to visit NATO to provide details of the assistance sought by the African Union. In June, following further consultations with the African Union, the European Union and the United Nations, NATO formally agreed to support the African Union with airlift and training.

On 1 July, The NATO airlift began and is coordinated from Europe. A special African Union Air Movement Cell at the African Union's headquarters in Addis Ababa, Ethiopia, coordinates the movement of incoming troops on the ground. Both the European Union and NATO are providing staff to support the cell, but the African Union has the lead.

NATO is also providing staff capacity building workshops for African Union officers within the Deployed Integrated Task Force Headquarters in Ethiopia. The training is based on strategic-level planning and focuses on technologies and techniques to create an overall analysis and understanding of Darfur and to identify the areas where the application of African Union assets can influence and shape the operating environment to deter crises. Following a request made by the African Union on 16 September, NATO decided to extend its assistance in the area of airlift and capacity-building until end March 2006.

Later, NATO-AU cooperation has mainly been pragmatic and driven by requests from the African Union for support in very specific areas. The principal areas of cooperation are: operational support, capacity-building support and support for the development of the African Union Standby Force. However, at the Warsaw Summit in 2016, NATO leaders committed to increasing political and practical cooperation with the African Union. At the same time, Allies also approved NATO's Framework for the South, which aims to integrate and streamline NATO's approach to tackle challenges by focusing on improved capabilities, enhanced anticipation and response, as well as boosting NATO's regional partnership and capacity-building efforts. Similarly, Allies agreed to the Projecting Stability initiative, a new vision to cooperate with partners beyond NATO territory, with an aim to develop a more strategic, coherent and effective approach to partnerships. More recently, in November 2019, NATO and the AU signed an agreement to strengthen political and practical partnership so as to better respond to common threats and challenges.

NATO's cooperation with the African Union is an integral part of both NATO's Framework for the South and the Alliance's efforts in Projecting Stability. Since the Warsaw Summit, NATO has strengthened its approach to the south and its partnerships in the region, and is continuing to develop relations with the AU.

From a practical point of view, Allied Joint Force Command (JFC) Naples is the NATO operational headquarters designated to implement the Alliance's practical cooperation with the AU.

JFC Naples is also the home of NATO's Strategic Direction South Hub, which was inaugurated in September 2017 as a way to face the current and evolving security issues from NATO's southern neighbourhood and enhance the Alliance's relationships with partners from the south.

Logistical support:

In January 2007, the AU made a general request to all partners, including NATO, for financial and logistical support to AMISOM. It later made a specific request to NATO in May 2007, requesting strategic airlift support for AU member states willing to deploy in Somalia under AMISOM. In June 2007, the North Atlantic Council (NAC) agreed, in principle, to support this request and NATO's support was initially authorised until August 2007. Strategic sealift support was requested at a later stage and agreed in principle by the NAC in September 2009.

The AU's strategic airlift and sealift support requests for AMISOM have been renewed on an annual basis. The current NAC agreement to support the AU with strategic air- and sealift for AMISOM extends until January 2021.

Planning support:

NATO provides subject matter experts for the AU Peace Support Operations Department. These experts have made significant contributions to AU priority areas. They have shared their knowledge and expertise in planning across various domains including maritime, finance, monitoring, procurement, air movement coordination, communications, information technology, logistics, human resources, military manpower management and contingencies. NATO's contribution of subject matter experts responds to annual requests from the AU. The areas requested vary from year to year based on AU priorities. In this capacity, NATO experts work side-by-side with AU counterparts, offering expertise in specific domains for periods of six to twelve months, renewable at the AU's request. The most recent request from the AU calls for support in strategic planning, as well as planning for movements and exercises.

Capacity-building support:

- Education and training:

NATO offers opportunities for AU personnel to attend courses at the NATO School in Oberammergau, Germany, the NATO Defence College in Rome, and other NATO training facilities such as NATO-accredited Centres of Excellence in the respective sponsoring countries. These education and training courses are offered based upon AU requirements and the availability of NATO training venues. On average, 20 AU students are sponsored at NATO training venues per year.

- Mobile training:

Since 2015 and in response to an AU request, NATO delivers dedicated training to African Union officers though Mobile Education and Training Teams (METT) that deliver tailored courses in Africa. NATO has progressively increased the number of courses delivered and, is providing three or more METT courses annually. The METT format allows for reach to a wider audience; participants are drawn from among AU staff, but also the Regional Economic Communities, which form the backbone of the development of Africa's continental force, the African Stand-by Force. On average, 30 AU students participate in each training session.

Support for the development of the African Standby Force:

NATO has been providing expert and training support to the African Standby Force (ASF) at the AU's request. The ASF is intended to be deployed in Africa in times of crisis and is part of the AU's efforts to develop long-term peacekeeping capabilities; it represents the AU's vision for a continental, on-call security apparatus, and shares similarities with the NATO Response Force.
At the AU's request, the Alliance offers capacity-building support through courses and training events. NATO has also organised certification/evaluation and training programmes

for AU staff, which support the ASF's operational readiness. For instance, NATO has trained AU officials participating in military exercises and provided military experts to assist in the evaluation and lessons learned procedures of an exercise. NATO has also supported various ASF preparatory workshops designed to develop ASF-related concepts. The Alliance is also specifically engaged in providing support to bringing the ASF's Continental Logistics base in Douala, Cameroon to full operational capacity.

NATO experts were also involved in supporting the preparation phases of Exercise Amani Africa II (October-November 2015) in South Africa, and played an active role in the execution phase. This was the first field training exercise for the ASF that brought together regional standby brigades from across the continent. African military, police and civilians participated in testing the ASF's rapid deployment capability and the ASF's level of readiness for full operational capability.

At the 2016 Summit in Warsaw, Allied leaders committed to expanding NATO's political and practical partnership with the AU to address common challenges. This has helped fuel a new momentum in NATO-AU relations to expand areas of cooperation. In April 2018, for instance, the NATO Defence College hosted a seminar in Rome, Italy, which convened senior officials from both NATO and the African Union in order to develop a series of pragmatic proposals to increase and enhance areas of cooperation. These include proposals in the areas of counter-terrorism, countering improvised explosive devices, the Women, Peace and Security agenda, building integrity, and support to AU peace-support operations.

Starting in 2005 with the provision of NATO logistical support to the AU to expand its mission in Darfur, the NATO-AU relationship has developed over time:

2005 – NATO provides strategic airlift to the African Union Mission in Sudan.

2007 – Allies agree to provide strategic airlift to support the AU's involvement in Somalia (AMISOM) and in 2009, agree to provide strategic sealift.

2011 – AU Commission Chairperson Jean Ping visits NATO twice in the context of Operation Unified Protector – the UN-mandated operation set up to protect civilians and civilian-populated areas, in Libya, under threat of attack.

2014 – AU Commissioner for Peace and Security Ambassador Smail Chergui visits NATO and signs the technical agreement on NATO-AU cooperation.

2015 - NATO opens its liaison office at AU headquarters in Addis Ababa.

2015 – NATO and the AU begin a programme of annual military-to-military staff talks.

2015 – NATO enhances the programme of mobile training solutions offered to AU officers.

2016 – NATO leaders agree to further strengthen and expand the Alliance's political and practical cooperation with the AU at the Warsaw Summit.

2019 – A cooperation agreement is signed to strengthen partnership and bring NATO and the AU closer together.

And to summarise, we can specify the operations to support peace in cooperation with the African Union carried out by the Alliance at the following points:

✓ NATO has developed cooperation with the African Union principally in three areas: operational support; capacity-building support; and assistance in developing and sustaining the African Standby Force (ASF).

✓ Operational support includes strategic air- and sealift, as well as planning support for the AU Mission in Somalia (AMISOM).

✓ Capacity-building support includes inviting AU officers to attend courses at NATO training and education facilities and delivering courses through NATO's Mobile Training Teams.

✓ Support for the development and sustainment of the ASF includes exercises and tailor-made training, as well as assistance in developing ASF-related concepts.

✓ NATO has also established a liaison office at AU headquarters in Addis Ababa, Ethiopia. It is led by a Senior Military Liaison Officer and provides, at AU's request, subject matter experts, who work in the AU's Peace and Security Department alongside African counterparts.

✓ NATO coordinates its AU-related work with bilateral partners and other international organisations, including the European Union and the United Nations.

## NATO and Libya

In February 2011, a peaceful protest in Benghazi in eastern Libya against the 42-year rule of Colonel Muammar Qadhafi met with violent repression, claiming the lives of dozens of protestors in a few days. As demonstrations spread beyond Benghazi, the number of victims grew. In response, the United Nations Security Council (UNSCR) adopted Resolution 1970 on 26 February 2011, which expressed "grave concern" over the situation in Libya and imposed an arms embargo on the country.

Following the adoption of Resolution 1970 and with growing international concern over the Libyan crisis, NATO stepped up its surveillance operations in the Mediterranean on 8 March 2011. The Alliance deployed Airborne Warning and Control Systems (AWACS) aircraft to the area to provide round-the-clock observation. These "eyes-in-the-sky" gave NATO detailed information about movements in Libyan airspace. Two days later the Alliance moved ships from current NATO assets, as well as ships made available by NATO nations for the mission, to the Mediterranean Sea to boost the monitoring effort.

After the situation in Libya further deteriorated, the UN Security Council adopted Resolution 1973 on 17 March 2011. The resolution condemned the "gross and systematic violation of human rights, including arbitrary detentions, enforced disappearances, torture and summary executions". It also introduced active measures, including a no-fly zone, and authorised member states, acting as appropriate through regional organisations, to use "all necessary measures" to protect Libyan civilians and civilian populated areas.

With the adoption of UNSCR 1973, several UN member states took immediate military action to protect civilians under Operation Odyssey Dawn. This operation, which was not under the command and control of NATO, was conducted by a multinational coalition led by the United States.

Responding to the United Nations' call:

On 22 March 2011, NATO responded to the UN's call to prevent the supply of "arms and related materials" to Libya by agreeing to launch an operation to enforce the arms embargo against the country. The next day, NATO ships operating in the Mediterranean began cutting off the flow of weapons and mercenaries to Libya by sea. NATO maritime assets stopped and searched any vessel they suspected of carrying arms, related materials or mercenaries to or from Libya.
In support of UNSCR 1973, NATO then agreed to enforce the UN-mandated no-fly zone over Libya on 24 March 2011.The resolution banned all flights into Libyan airspace to protect civilian-populated areas from air attacks, with the exception of flights used for humanitarian and aid purposes.

The Alliance took sole command and control of the international military effort for Libya on 31 March 2011. NATO air and sea assets began to take military actions to protect civilians and civilian populated areas. Throughout the crisis, the Alliance consulted closely with the UN, the League of Arab States and other international partners.

Commitment to protecting the Libyan people:

The Alliance's decision to undertake military action was based on three clear principles: a sound legal basis, strong regional support and a demonstrable need. By the end of March 2011, OUP had three distinct components:

- Enforcing an arms embargo in the Mediterranean Sea to prevent the transfer of arms, related materials and mercenaries to Libya.

- Enforcing a no-fly zone to prevent aircrafts from bombing civilian targets.

- Conducting air and naval strikes against military forces involved in attacks or threatening to attack Libyan civilians and civilian populated areas.

During a meeting in Berlin on 14 April 2011, foreign ministers from NATO Allies and non-NATO partners agreed to continue OUP until all attacks on civilians and civilian populated areas ended, the Qadhafi regime withdrew all military and para-military forces to bases, and the regime permitted immediate, full, safe and unhindered access to humanitarian aid for the Libyan people.

On 8 June 2011, NATO defence ministers met in Brussels and agreed to keep pressure on the Qadhafi regime for as long as it took to end the crisis, reaffirming the goals laid out by the foreign ministers.

Following the liberation of Tripoli on 22 August by opposition forces, the Secretary General reaffirmed both NATO's commitment to protect the Libyan people and its desire that the Libyan people decide their future in freedom and in peace.

International heads of state and government further reiterated this commitment during a "Friends of Libya" meeting in Paris on 1 September.

On 16 September, the UN Security Council adopted Resolution 2009, which unanimously reasserted NATO's mandate to protect civilians in Libya. The new resolution also established a United Nations Support Mission in Libya (UNSMIL).

Ending the mission:

As NATO air strikes helped to gradually degrade the Qadhafi regime's ability to target civilians, NATO defence ministers met in Brussels on 6 October and discussed the prospects of ending OUP. Ministers confirmed their commitment to protect the people of Libya for as long as threats persisted, but to end the mission as soon as conditions permitted. The NATO Secretary General also pledged to coordinate the termination of operations with the UN and the new Libyan authorities.

A day after opposition forces captured the last Qadhafi regime stronghold of Sirte and the death of Colonel Qadhafi on 20 October 2011, the North Atlantic Council took the preliminary decision to end OUP at the end of the month. During that transition period, NATO continued to monitor the situation and retained the capacity to respond to threats to civilians, if needed.

A week later, the North Atlantic Council confirmed the decision to end OUP. On 31 October 2011 at midnight Libyan time, a NATO AWACS concluded the last sortie; 222 days after the operation began. The next day, NATO maritime assets left Libyan waters for their home ports. Although NATO's operational role regarding Libya is finished, the Alliance stands ready to assist Libya in areas where it could provide added value, such as in the area of defence and security sector reforms, if requested to do so by the new Libyan authorities.


**Counter-piracy operations**

High levels of piracy activity in the Gulf of Aden, off the Horn of Africa and in the Indian Ocean undermined international humanitarian efforts in Africa and the safety of one of the busiest and most important maritime routes in the world – the gateway in and out of the Suez Canal – for a long time. Between 2008 and 2016, NATO helped to deter and disrupt pirate attacks, while protecting vessels and helping to increase the general level of security in the region through different military operations.

We can summarise the actions of the Alliance in counter-piracy operations in the following points:

- ✓ In 2008, at the request of the United Nations, NATO started to support international efforts to combat piracy in the Gulf of Aden, off the Horn of Africa and in the Indian Ocean with Operation Allied Provider and Allied Protector.

✓ From August 2009, NATO then led Operation Ocean Shield, which helped to deter and disrupt pirate attacks, while protecting vessels and helping to increase the general level of security in the region.

✓ NATO worked in close cooperation with other actors in the region including the European Union's Operation Atalanta, the US-led Combined Task Force 151 and individual country contributors.

✓ The very presence of this international naval force deterred pirates from pursuing their activities and contributed to the suppression of piracy in the region. The implementation of best management practices by the shipping industry, as well as the embarkation of armed security teams on board, also contributed to this trend.

✓ With no successful piracy attacks since 2012, NATO terminated Ocean Shield on 15 December 2016. However, NATO is remaining engaged in the fight against piracy by maintaining maritime situational awareness and continuing close links with other international counter-piracy actors.

✓ NATO is also maintaining its counter-piracy efforts at sea and ashore, by supporting countries in the region to build the capacity to fight piracy themselves.

# THE ALLIANCE'S ROLE IN FIGHTING NEW THREATS

## The NATO's role in the fight against terrorism

We all bear in mind that the terrorist attacks on New York and Washington, DC of 11 September 2001 thrust not only the United States but the entire Alliance into the fight against terrorism. Less than 24 hours after the attacks, NATO invoked Article 5 (the collective defence clause of its founding treaty) for the very first time in its history. The political significance of this decision resides in the fact that Article 5 involves a commitment by each of the Allies to consider an attack on one or more of them in Europe or North America as an attack against them all. As a consequence, these attacks were considered an attack on all the members of the Alliance, and member and Partner countries alike firmly and repeatedly condemned the attacks and terrorism in all its forms.

This way, the practical implications of the decision were unprecedented since it was the first time that the Alliance deployed forces and other assets in support of an Article 5 operation. At the request of the United States, the Allies agreed to take eight specific measures of support. One of these was to send NATO Airborne Warning and Control System (AWACS) aircraft to the United States to assist in patrolling American airspace. The operation was known as Eagle Assist and ran until mid-May 2002. Another was the launch, on 26 October 2001, of a counter-terrorist operation in the Mediterranean called Active Endeavour, which it was terminated in October 2016 and succeeded by Sea Guardian.

Therefore, the invocation of Article 5 and subsequent operations to help protect US airspace and patrol the Mediterranean were followed by another very important first operation for NATO outside of Alliance territory. The Alliance conducted its first peacekeeping operation outside Europe when it decided, in August 2003, to take over the International Security Assistance Force (ISAF) in Afghanistan. This was later followed by other out-of-area missions, as we have previously seen.

At the same time, a major review of military capabilities was underway with the launch of the Prague Capabilities Commitment and the decision to create a NATO Response Force (NRF).

Furthermore, together with the contribution to the fight against terrorism, existing NATO operations also took on a role in this area. NATO introduced a number of political initiatives and practical measures in many different areas to help combat terrorism. It adopted a Military Concept for Defence against Terrorism, reinforced cooperation with Partner countries by agreeing on a Partnership Action Plan against Terrorism and introduced measures against the spread of weapons of mass destruction. An enhanced package of anti-terrorist measures was adopted and new initiatives were introduced to improve cyber-defence, civil emergency planning and civil protection. These measures were bolstered by the commitment to reinforce cooperation with other international organisations on terrorism.

Initial support to the United States:

At the request of the United States, on 4 October 2001, NATO Allies agreed to take eight measures to expand the options available in the campaign against terrorism. These eight measures included the following:

- greater intelligence sharing;

- assistance to states threatened as a result of their support for coalition efforts;

- increased security for the United States, and other Allies' facilities on their territory;

- back-filling of selected Allied assets needed to support anti-terrorist operations;

- blanket overflight rights for the United States' and other Allies' aircraft for military flights related to counter-terrorism operations;

- access to ports and airfields;

- the deployment of NATO naval forces to the eastern Mediterranean;

- the deployment of elements of the NATO' Airborne Early Warning and Control Force to support counter-terrorism operations.

As we have seen previously, Operation Eagle Assist was terminated by the North Atlantic Council in May 2002 following material upgrades to the US air defence posture, enhanced cooperation between US civil and military authorities, and a US re-evaluation of homeland security requirements.

Operation Active Endeavour:

As we have also seen, since October 2001, elements of NATO's Standing Naval Forces have conducted anti-terrorist operations in the Mediterranean. Known as Operation Active Endeavour, the operation has made use of ships, submarines and aircraft, initially to monitor merchant shipping in the eastern Mediterranean. The mission was expanded in March 2003 to include escorting non-military ships from Alliance member countries through the Straits of Gibraltar, and again in April 2003 to include compliant boarding of suspicious vessels in accordance with the rules of international law. A year later, the operation was extended to the entire Mediterranean.

Besides, since the beginning of their operational role in 2001, forces assigned to the operation have hailed over 70 000 merchant vessels in the Mediterranean, conducted surveillance operations using ships, submarines and aircraft in order to provide an overview of maritime activity in the area, boarded approximately 100 vessels in accordance with the rules of international law, and escorted several hundred vessels through the Straits of Gibraltar. In addition to NATO's Standing Naval Forces, to which a number of member countries contribute, American and Portuguese maritime patrol and other aircraft, Spanish aircraft, helicopters and frigates, and Danish, Norwegian and German patrol boats have participated in these operations.

On the other hand, with the extension of its area of operations to the entire Mediterranean in 2003, the scope of potential multinational support for Operation Active Endeavour has also been widened to include NATO's Partner countries and Mediterranean Dialogue countries.

Russian and Ukrainian offers to contribute to the operation, in the framework of NATO-Russian and NATO-Ukrainian cooperative arrangements, were also welcomed by Alliance leaders.

We have to mention that during their deployment in the Mediterranean, the forces involved in the operation have been called upon on several occasions to participate in emergency operations involving, for example, the evacuation of oil rig personnel threatened by high winds and heavy seas and the rescue of passengers aboard a ferry.

International Security Assistance Force in Afghanistan:

Remember that military operations led by the United States in Afghanistan resulted in the ousting of the Taliban regime, its replacement by an administration committed to peace and to rebuilding the country, and the disabling of large parts of the extensive al-Qaida network in Afghanistan and elsewhere. Known as Operation Enduring Freedom, this effort has been supported by a number of NATO countries that have, for example, provided special forces teams or contributed planes and ships to work together with US special forces on surveillance, interdiction and interception operations. Offers of support have also been made by a number of other non-NATO countries, including Russia and Ukraine.

At the same time, NATO forces have played a crucial role in the UN-mandated multinational force initially led by individual NATO countries, the International Security Assistance Force (ISAF), which was transferred to unified NATO command in August 2003. The role of the force is to help stabilise the country and create the conditions for self-sustaining peace. In this respect, ISAF can be considered as part of NATO's fight against terrorism since it is helping, albeit indirectly, to put an end to terrorist activity on Afghan territory.

Let us also remember that ISAF is a multinational force drawn from NATO and Partner countries. Initially under UK command, the force was under Turkish command from June 2002 and, from February 2003, under the joint command of Germany and the Netherlands, with NATO support in specific fields. Examples of national contributions included airlift capability provided by Belgium, a field hospital provided by the Czech Republic, a medical team contributed by Portugal and engineering and logistical support provided by Poland.

On 17 October 2002, a request from Germany and the Netherlands for NATO support in preparing for this role was approved by the North Atlantic Council. NATO assistance was sought in particular in the areas of force generation, communications, and intelligence coordination and information sharing.

A force generation conference attended by participants from NATO and Partner countries was held at the Supreme Headquarters Allied Powers in Europe (SHAPE) on 27 November 2002 to give the countries an opportunity to make offers of contributions and to identify and discuss critical shortfalls that might need to be filled to enhance future capacity. This was the first such conference to take place in support of countries offering to lead a non-NATO-led military operation based on a United Nations Security Council resolution.

Anti-terrorist operations in the Balkans:

Equally remarkable is that NATO operations in the Balkans have contributed to making that region less prone to terrorist activities. Action has been taken by NATO-led forces against local terrorist groups with links to the al-Qaida network, as part of the wider campaign against terrorism, particularly through measures aimed at curtailing illegal movements of people, arms and drugs.

Military Concept for Defence against Terrorism:

A Military Concept for Defence against Terrorism was approved at the November 2002 Prague Summit. It underlines the Alliance's readiness to help deter, defend, disrupt and protect against terrorist attacks or the threat of such attacks directed from abroad against Allied populations, territory, infrastructure and forces, including by acting against terrorists and those who harbour them; to provide assistance to national authorities in dealing with the consequences of terrorist attacks; to support operations by the European Union or other international organisations or coalitions involving Allies; and to deploy forces as and where required to carry out such missions.

The Partnership Action Plan on Terrorism:

Together with its Partner countries, NATO has elaborated a Partnership Action Plan on Terrorism (PAP-T). Issued at the Prague Summit in November 2002, the PAP-T provides a framework for cooperation on terrorism and defines Partnership roles and the instruments for fighting terrorism and managing its consequences. Mediterranean Dialogue countries can also participate in activities under the plan.

The Euro-Atlantic Partnership Council, the NATO-Russia Permanent Joint Council (succeeded by the NATO-Russia Council in May 2002), the NATO-Ukraine Commission and countries participating in NATO's Mediterranean Dialogue all joined NATO in condemning the September 11 attacks and offering their support to the United States. NATO countries continue to make extensive use of Partnership mechanisms to consult with their Partner countries about further steps. They agree that a comprehensive effort comprising political, economic, diplomatic and military actions as well as law enforcement measures is needed to combat terrorism; in other words, a long-term, multifaceted approach involving NATO as a whole but also involving all the Allies individually, both as members of the Alliance and as members of the United Nations, the Organization for Security and Co-operation in Europe (OSCE) and the European Union.

An enhanced package of anti-terrorist measures:

Equally, an enhanced package of anti-terrorist measures was agreed at the Istanbul Summit in June 2004. These measures include improved intelligence sharing through NATO's Terrorist Threat Intelligence Unit and other means; improving NATO's ability to respond rapidly to national requests for assistance in response to a terrorist attack; helping to provide protection during selected major events, including the use of NATO airborne early warning aircraft; strengthening the contribution of NATO-led operations in the Mediterranean, the Balkans and Afghanistan to the fight against terrorism, increasing cooperation with Partner countries and with other international and regional organisations; and improving relevant capabilities. They also include a specialised armaments programme endorsed by the Conference of National Armaments Directors at its meeting in May 2004. This programme focuses on ten areas:

- actions to counter improvised explosive devices, such as car and roadside bombs;

- reduction of the vulnerability of wide-body civilian and military aircraft to man-portable air defence missiles;

- reduction of the vulnerability of helicopters to rocket-propelled grenades;

- protection of harbours and ships from explosive-packed speedboats and underwater divers;

- detection, protection and defeat of chemical, biological, radiological and nuclear weapons;

- explosive ordnance disposal;

- precision airdrop technology for special operations forces and their equipment; - intelligence, surveillance, target acquisition and reconnaissance of terrorists; - technologies to counter mortar attacks;

- protection of critical infrastructure.

**Countering terrorism today**

Terrorism in all its forms poses a direct threat to the security of the citizens of NATO countries, and to international stability and prosperity. It is a persistent global threat that knows no border, nationality or religion and is a challenge that the international community must tackle together. NATO will continue to fight this threat in all its forms and manifestations with determination and in full solidarity. NATO's work on counter-terrorism focuses on improving awareness of the threat, developing capabilities to prepare and respond, and enhancing engagement with partner countries and other international actors.

Awareness:

In support of national authorities, NATO ensures shared awareness of the terrorist threat through consultations, enhanced intelligence-sharing and continuous strategic analysis and assessment.

Intelligence reporting at NATO is based on contributions from Allies' intelligence services, both internal and external, civilian and military. The way NATO handles sensitive information has gradually evolved, based on successive summit decisions and continuing reform of intelligence structures since 2010. Since 2017, the Joint Intelligence and Security Division at NATO benefits from increased sharing of intelligence between member services and the Alliance, and produces strategic analytical reports relating to terrorism and its links with other transnational threats.

Intelligence-sharing between NATO and partner countries' agencies continues through the Intelligence Liaison Unit at NATO Headquarters in Brussels, and an intelligence liaison cell at Allied Command Operations (ACO) in Mons, Belgium. An intelligence cell at NATO Headquarters improves how NATO shares intelligence, including on foreign fighters. NATO faces a range of threats arising from instability in the region to the south of the Alliance. NATO increases its understanding of these challenges and improves its ability to respond to them through the "Hub for the South" based at NATO's Joint Force Command in Naples, Italy. The Hub collects and analyses information, assesses potential threats and engages with partner nations and organisations.

Beyond the everyday consultations within the Alliance, experts from a range of backgrounds are invited to brief Allies on specific areas of counter-terrorism. Likewise, discussions with international organisations, including the United Nations (UN), the European Union (EU), the Organization for Security and Co-operation in Europe (OSCE) and the Global Counterterrorism Forum (GCTF), enhance Allies' knowledge of international counter-terrorism efforts worldwide and help NATO refine the contribution that it makes to the global approach.

Capabilities:

The Alliance strives to ensure that it has adequate capabilities to prevent, protect against and respond to terrorist threats. Capability development and work on innovative technologies are part of NATO's core business, and methods that address asymmetric threats including terrorism and the use of non-conventional weapons, are of particular relevance. Much of this work is conducted through the Defence Against Terrorism Program of Work (DAT POW), which aims to protect troops, civilians and critical infrastructure against attacks perpetrated by terrorists, such as suicide attacks, improvised explosive devices (IEDs), rocket attacks against Aircraft and helicopters and attacks using chemical, biological or radiological material.

NATO's Centers of Excellence are important contributors to many projects, providing expertise across a range of topics including military engineering for route clearance, countering IEDs, explosives disposal, cultural familiarisation, network analysis and modeling.

a) Defence Against Terrorism Program of Work:

- The DAT POW was developed by the Conference of National Armaments Directors (CNAD) in 2004. Its primary focus was on technological solutions to mitigate the effects of terrorist attacks but the program has since widened its scope to support comprehensive capability development. It now includes exercises, trials, development of prototypes and concepts, and interoperability demonstrations. Most projects under the program focus on finding solutions that can be fielded in the short term and that respond to the military needs of the Alliance. The program uses new or adapted technologies or methods to detect, disrupt and defeat asymmetric threats under three capability umbrellas: incident management, force protection / survivability, and network engagement.

b) Countering chemical, biological, radiological and nuclear threats:

- The spread and potential use of weapons of mass destruction (WMD) and their delivery systems together with the possibility that terrorists will acquire them, are acknowledged as priority threats to the Alliance. Therefore, NATO places a high priority on preventing the proliferation of WMD to state and non-state actors and defending against chemical, biological, radiological and nuclear (CBRN) threats and hazards that may pose a threat to the safety and security of Allied populations. The NATO Combined Joint CBRN Defence Task Force is designed to respond to and manage the consequences of the use of CBRN agents. The NATO-certified Center of Excellence on Joint CBRN Defence, in the Czech Republic, further enhances NATO's capabilities.

c) Countering terrorist misuse of technology:

- Terrorists have sought to use and manipulate various technologies in their operations, including easily available off-the-shelf technology. Drones in particular have been identified as a threat. Therefore, in February 2019, defence ministers agreed to a practical framework to counter unmanned aerial systems (UAS). A program of work will be implemented over the next two years, helping to coordinate approaches and identify additional steps to address this threat.

d) Operations:

- Since 2017, NATO has been a member of the Global Coalition to Defeat ISIS. As a member of the Coalition, NATO has been playing a key role in the fight against international terrorism for many years, including through its long-standing operational engagement in Afghanistan, through intelligence-sharing, and through its work with partners with a view to projecting stability in the Euro-Atlantic area and beyond.

- At the 2016 Summit in Warsaw, Allied leaders agreed to provide direct support to the Coalition through the provision of NATO AWACS surveillance aircraft. The first patrols of NATO AWACS aircraft, operating from Konya Airfield in Turkey, started in October 2016.

- Moreover, the Alliance decided to launch a training and capacity-building activity to train, advise and assist Iraqi forces both in Iraq and Jordan. Finally, in December 2018, Allies updated an Action Plan to fight terrorism, which included NATO's membership in the Coalition as well as more AWACS flight time and information-sharing. Following a request by the Iraqi government and the Coalition, Allies agreed in February 2018 on planning and direction for a NATO Training and Capacity Building Mission in Iraq, with the aim of making the Alliance's efforts more sustainable and taking on additional tasks as required. At the 2018 Brussels Summit, Allies agreed to launch this non-combat training and capacity-building mission in Iraq.

- Also in the summer of 2018, Allies agreed a new biometric data policy, consistent with applicable national and international law and subject to national requirements and restrictions. The policy enables biometric collection to support NATO operations, based upon a mandate from the North Atlantic Council. The policy is particularly relevant to the threat posed by foreign terrorist fighters. United Nations Security Council Resolution (UNSCR) 2396 highlights the acute and growing threat posed by foreign terrorist fighters and "urges Member States to expeditiously exchange information, through bilateral or multilateral mechanisms and in accordance with domestic and international law, concerning the identity of Foreign Terrorist Fighters". NATO works to maintain its military capacity for crisis management and humanitarian assistance operations. When force deployment is necessary, counter-terrorism considerations are often relevant. Lessons learned in operations, including by Special Operations Forces, must not be wasted. Interoperability is essential if members of future coalitions are to work together. Best practices are, therefore, incorporated into education, training and exercises.

- In December 2019, Allied Leaders underlined their commitment to the fight against terrorism and agreed an updated Action Plan to step up the Alliance's efforts in the fight against terrorism. Allies remain also committed to supporting the Global Coalition to Defeat ISIS as well as to NATO's training missions in Iraq and Afghanistan.

- NATO's maritime operation Sea Guardian is a flexible maritime security operation that is able to perform the full range of maritime security tasks, including countering terrorism at sea if required. It succeeded operation "Active Endeavor", which was launched in 2001 under Article 5 of NATO's founding treaty as part of NATO's immediate response to the 9/11 terrorist attacks to deter, detect and, if necessary, disrupt the threat of terrorism in the Mediterranean Be. Active Endeavor was terminated in October 2016.

- Many other operations have had relevance to international counter-terrorism efforts. For example, the International Security Assistance Force (ISAF) (the NATO-led operation in Afghanistan, which began in 2003 and came to an end in 2014) helped the government expand its authority and implement security to prevent the country once again becoming a safe haven for international terrorism. Resolute Support Mission in Afghanistan, which followed ISAF and stood up in 2015, is a non-combat mission that builds capacity in the Afghan security forces.

e) Crisis management:

- NATO's long-standing work on civil emergency planning, critical infrastructure protection and crisis management provides a resource that may serve both Allies and partners upon request. This field can relate directly to counter-terrorism, building resilience and ensuring appropriate planning and preparation for response to and recovery from terrorist acts.

f) Protecting populations and critical infrastructure

- National authorities are primarily responsible for protecting their population and critical infrastructure against the consequences of terrorist attacks, CBRN incidents and natural disasters. NATO can assist nations by developing non-binding advice and minimum standards and act as a forum to exchange best practices and lessons learned to improve preparedness and national resilience. NATO has developed Guidelines for first response to a CBRN incident as well as for civil-military cooperation in case of a CBRN terrorist attack and organisations "International Courses for Trainers of First Responders to CBRN Incidents". NATO guidance can also advise national authorities on warning the general public and alerting emergency responders. NATO can call on an extensive network of civil experts, from government and industry, to help respond to requests for assistance. Its Euro-Atlantic Disaster Response Coordination Center (EADRCC) coordinates responses to national requests for assistance following natural and man-made disasters including terrorist acts involving CBRN agents.

Engagement:

As the global counter-terrorism effort requires a holistic approach, Allies have resolved to strengthen outreach to and cooperation with partner countries and international actors.

1. With partners:

- Increasingly, partners are taking advantage of partnership mechanisms for dialogue and practical cooperation relevant to counter-terrorism, including defence capacity building. Interested partners are encouraged to include a section on counter-terrorism in their individual cooperation agreements with NATO. Allies place particular emphasis on shared awareness, capacity building, civil emergency planning and crisis management to enable partners to identify and protect vulnerabilities and to prepare to fight terrorism more effectively. Countering improvised explosive devices, the promotion of a whole-of-government approach and military border security are among NATO's areas of work with partners.

- Counter-terrorism is one of the five priorities of the NATO Science for Peace and Security (SPS) Program. The SPS Program enhances cooperation and dialogue between scientists and experts from Allies and partners, contributing to a better understanding of the terrorist threat, the development of detection and response measures, and fostering a network of experts.

- Activities include workshops, training courses and multi-year research and development projects that contribute to identifying: methods for the protection of critical infrastructure, supplies and personnel; human factors in defence against terrorism; technologies to detect explosive devices and illicit activities; and risk management, best practices, and use of new technologies in response to terrorism. The SPS Program is flexible and able to respond to evolving priorities. As examples, in 2017 SPS issued to Call for Proposals to address human, social, cultural, scientific and technological advancements in the fight against terrorism. In 2018, NATO launched a new initiative to develop an integrated system of sensors and data fusion technologies capable to detect explosives and concealed weapons in real time and to secure mass transport infrastructures, such as airports, metro and railway stations.

- On 1 April 2014, Allied foreign ministers condemned Russia's illegal military intervention in Ukraine and Russia's violation of Ukraine's sovereignty and territorial integrity. Ministers underlined that NATO does not recognise Russia's illegal and illegitimate attempt to annex Crimea. As a result, ministers decided to suspend all practical civilian and military cooperation between NATO and Russia, including in the area of counter-terrorism, which had been among the main drivers behind the creation of the NATO-Russia Council in May 2002. This decision was reconfirmed by Allied leaders at the Wales Summit in September 2014 and to date, practical cooperation with Russia remains suspended.

2. With international actors:

- NATO cooperates in particular with the UN, the EU and the OSCE to ensure that views and information are shared and that appropriate action can be taken more effectively in the fight against terrorism. The UN Global Counter-Terrorism Strategy, international conventions and protocols against terrorism, together with relevant UN resolutions provide a common framework for efforts to combat terrorism.

- NATO works closely with the UN Counter-Terrorism Committee and its Executive Directorate as well as with the Counter-Terrorism Implementation Task Force and many of its component organisations, including the UN Office on Drugs and Crime. NATO's Centers of Excellence and education and training opportunities are often relevant to UN counter-terrorism priorities, as is the specific area of explosives management. More broadly, NATO works closely with the UN agencies that play a leading role in responding to international disasters and in consequence management, including the UN Office for the Coordination of Humanitarian Affairs, the Organization for the Prohibition of Chemical Weapons and the UN 1540 Committee. In March 2019, NATO and the UN launched a joint project to improve CBRN resilience in Jordan.

- NATO and the European Union are committed to combatting terrorism and the proliferation of weapons of mass destruction. They exchange information regularly on counter-terrorism projects and on related activities such as work on the protection of civilian populations against chemical, biological, radiological and nuclear (CBRN) attacks. Relations with the European External Action Service's Counter-terrorism section, with the Counter-terrorism Coordinator's office and other parts of the EU help ensure mutual understanding and complementarity.

- NATO maintains close relations with the OSCE's Transnational Threats Department's Action against Terrorism Unit and with field offices and the Border College in Dushanbe (Tajikistan), which works to create secure open borders through specialised training of senior officers from national border security agencies.

- NATO is also working with other regional organisations to address the terrorism threat. In April 2019, NATO and the African Union held their first joint counter-terrorism training in Algiers.

- The use of civilian aircraft as a weapon in the 9/11 terrorist attacks led to efforts to enhance aviation security. NATO contributed to improved civil-military coordination of air traffic control by working with EUROCONTROL, the International Civil Aviation Organization (ICAO), the US Federal Aviation Administration, major national aviation and security authorities, airlines and pilot associations and the International Air Transport Association ( IATA).

3. Education:

- NATO offers a range of training and education opportunities in the field of counter-terrorism to both Allies and partner countries. It can draw on a wide network that includes the NATO School in Oberammergau, Germany, mobile training courses run out of Joint Force Commands at Naples and Brunssum and the Centers of Excellence (COEs) that support the NATO Command Structure. There are more than 20 COEs fully accredited by NATO of which several have a link to the fight against terrorism. The Center of Excellence for Defence Against Terrorism (COE-DAT) in Ankara, Turkey serves both as a location for meetings and as a catalyst for international dialogue and discussion on terrorism and counter-terrorism. The COE-DAT reaches out to over 50 countries and 40 organisations.

## The NATO's role in the fight against proliferation of weapons of mass destruction

The proliferation of nuclear weapons and other weapons of mass destruction (WMD), and their delivery systems, could have incalculable consequences for national, regional and global security. During the next decade, proliferation will remain most acute in some of the world's most volatile regions. The potential effects of WMD proliferation on NATO Allies are one of the greatest threats NATO faces.

Protecting against weapons of mass destruction:

NATO governments endorsed the implementation of five nuclear, biological and chemical (NBC) defence initiatives designed to improve the Alliance's defence capabilities against weapons of mass destruction at the Prague Summit in November 2002. These consist of:

• a prototype deployable NBC analytical laboratory;

• an NBC joint assessment team;

• a virtual centre of excellence for NBC weapons defence;

• a NATO biological and chemical defence stockpile; and

• a disease surveillance system.

NATO Allies engage in preventing the proliferation of WMD by state and non-state actors through an active political agenda of arms control, disarmament and non-proliferation. They also do this by developing and harmonising defence capabilities and, when necessary, by employing these capabilities, consistent with political decisions in support of non-proliferation objectives. Both political and defence elements are essential to NATO's security.

Since the launch of the 1999 WMD Initiative, which was designed to integrate political and military aspects of NATO work in responding to WMD proliferation, Allies have continued to intensify and expand NATO's contribution to global non-proliferation efforts. Through cooperation with partners and relevant international organisations, NATO has historically provided strong support to the negotiations and implementation of a number of arms control and non-proliferation regimes.

Allies have also intensified NATO's defence response to the risk posed by WMD by improving civil preparedness and consequence-management capabilities in the event of WMD use or a chemical, biological, radiological or nuclear (CBRN) accident or incident.

The Arms Control, Disarmament, and WMD Non-proliferation Centre (ACDC):

The ACDC was created in 2017, merging NATO's Arms Control and Coordination Section with the WMD Non-Proliferation Centre. The ACDC resides in the Political Affairs and Security Policy Division at NATO Headquarters and comprises national experts as well as personnel from NATO's International Staff and International Military Staff.

Improving CBRN defence capabilities:

NATO continues to significantly improve its CBRN defence posture with the establishment of the Combined Joint CBRN Defence Task Force (CJ-CBRND-TF), the NATO CBRN Reachback capability, the Joint CBRN Defence Centre of Excellence (JCBRN Defence COE), the Defence against Terrorism Centre of Excellence (COE), and other COEs and agencies that support NATO's response to the WMD threat. Allies continue to invest significant resources in capabilities ranging from CBRN reconnaissance and decontamination to warning and reporting, individual protection, and CBRN hazard management.

Combined Joint CBRN Defence Task Force:

The NATO Combined Joint CBRN Defence Task Force is designed to perform a full range of CBRN defence missions. It comprises the multinational CBRN Defence Battalion and the Joint Assessment Team.

The Task Force is led by an individual Ally on a 12-month rotational basis. Under normal circumstances, it operates within the NATO Response Force (NRF), which is a multinational force designed to respond rapidly to emerging crises across the full spectrum of Alliance missions.

However, the Task Force may operate independently of the NRF on other tasks as required, for example, helping civilian authorities in NATO member countries.

Joint Centre of Excellence on CBRN Defence:

The JCBRN Defence COE in Vyškov, Czech Republic, was activated in July 2007. It is an international military organisation sponsored and manned by the Czech Republic, France, Germany, Greece, Hungary, Italy, Poland, Romania, Slovakia, Slovenia, the United Kingdom and the United States. It is also open for partners that want to become contributing nations. Austria joined the Centre as the first such contributing nation in 2016.

The COE offers recognised expertise and experience in the field of CBRN to the benefit of the Alliance. It provides opportunities to improve interoperability and capabilities by enhancing multinational education, training and exercises; assisting in concept, doctrine, procedures and standards development; and testing and validating concepts through experimentation. It has thus supported NATO's transformation process.

The COE integrates a CBRN Reachback Element (RBE), which has reached Full Operational Capability (FOC) in January 2016. This Reachback capability provides timely and comprehensive scientific (technical) and operational CBRN expertise, assessments and advice to NATO commanders, their staff and deployed forces during planning and execution of operations. The RBE, together with its secondary network which comprises various civilian and military institutions, is able, if needed, to operate 24/7.

Standardisation, training, research and development:

NATO creates and improves necessary standardisation documents, conducts training and exercises, and develops necessary capability improvements in the field of CBRN defence through the work of many groups, bodies and institutions, including:

- CBRN Medical Working Group;

- Joint CBRN Defence Capability Development Group;

- NATO Research and Technology Organization; and

- Partnerships and Cooperative Security Committee (taking over the task of developing and implementing science activities, which were formerly managed under the auspices of the Science for Peace and Security Committee).

The Alliance also continues to create and improve standard NATO agreements that govern Allied operations in a CBRN environment. These agreements guide all aspects of preparation, ranging from standards for disease surveillance to rules for restricting troop movements. In addition, the Organization conducts training exercises and senior-level seminars that are designed to test interoperability and prepare NATO leaders and forces for operations in a CBRN environment.

Building capacity and scientific collaboration:

The NATO Science for Peace and Security (SPS) Programme enables collaboration between NATO and partner countries on issues of common interest to enhance their mutual security by facilitating international research efforts to meet emerging security challenges, supporting

NATO-led operations and missions, and advancing early warning and forecast for the prevention of disasters and crises.

The central objective of SPS activities in WMD non-proliferation and CBRN defence is to improve the ability of NATO and its partners to protect their populations and forces from CBRN threats. The Programme supports research towards the development of CBRN defence capabilities, training activities and workshops in the following fields:

1. protection against CBRN agents, as well as diagnosing their effects, detection, decontamination, destruction, disposal and containment;

2. risk management and recovery strategies and technologies; and

3. medical counter-measures for CBRN agents.

Arms control, disarmament and non-proliferation:

Arms control, disarmament and non-proliferation are essential tools in preventing the use of WMD and the spread of these weapons and their delivery systems. That is why Allies will continue to support numerous efforts in the fields mentioned above, always based on the principle to ensure undiminished security for all Alliance members.

Since the end of the Cold War, Allies have dramatically reduced the number of nuclear weapons stationed in Europe and their reliance on nuclear weapons in the NATO strategy. No NATO member country has a chemical or biological weapons programme. Additionally, Allies are committed to destroying stockpiles of chemical agents and have supported a number of partners and other countries in this work.

NATO members are resolved to seek a safer world for all and create the conditions for a world without nuclear weapons in accordance with the goal of the Nuclear Non-Proliferation Treaty (NPT). That is why the Alliance will seek to create the conditions for further reductions in the future. One important step towards this goal is the implementation of the New Strategic Arms Reduction Treaty (START) between the United States and the Russian Federation.

With respect to the new Treaty on the Prohibition of Nuclear Weapons, the North Atlantic Council declared that the treaty disregards the realities of the increasingly challenging international security environment. At a time when the world needs to remain united in the face of growing threats, in particular the grave threat posed by North Korea's nuclear programme, the treaty fails to take into account these urgent security challenges. This new treaty risks undermining the NPT, which has been at the heart of global non-proliferation and disarmament efforts for almost 50 years, and the International Atomic Energy Agency (IAEA) Safeguards regime which supports it. In view of this and a number of other arguments including their commitment to advancing security through deterrence, defence, disarmament, non-proliferation and arms control, the Allied nations cannot support this treaty.

Improving civil preparedness:

National authorities are primarily responsible for protecting their populations and critical infrastructure against the consequences of terrorist attacks, CBRN incidents and natural disasters. Within NATO, Allies have agreed baseline requirements for national resilience and are developing guidelines to help nations achieve them. The Alliance also serves as a forum to exchange best practices and lessons learned to improve preparedness and national resilience.

A network of 380 civil experts from across the Euro-Atlantic area exists to support these efforts. Their expertise covers all civil aspects relevant to NATO planning and operations, including crisis management, consequence management and critical infrastructure protection. Drawn from government and industry, experts participate in training and exercises, and respond to requests for assistance.

Under the auspices of NATO's Euro-Atlantic Disaster Response Coordination Centre (EADRCC), Allies have established an inventory of national civil and military capabilities that could be made available to assist stricken countries following a CBRN terrorist attack. Originally created in 1998 to coordinate responses to natural and man-made disasters, the EADRCC has since 2001 been given an additional coordinating role for responses to potential terrorist acts involving CBRN agents. It organises major international field exercises to practise responses to simulated disaster situations and consequence management.

Cooperating with partners:

The Alliance engages actively to enhance international security through partnership with relevant countries and other international organisations. NATO's partnership programmes are therefore designed as a tool to provide effective frameworks for dialogue, consultation and coordination. They contribute actively to NATO's arms control, non-proliferation and disarmament efforts.

Examples of institutionalised fora of the aforementioned cooperation include the Euro-Atlantic Partnership Council, the NATO-Ukraine Commission, the NATO-Georgia Commission and the Mediterranean Dialogue. NATO also consults with countries in the broader Middle East region which take part in the Istanbul Cooperation Initiative, as well as with partners across the globe.

International outreach activities:

Outreach to partners, international and regional organisations helps develop a common understanding of the WMD threat and encourages participation in and compliance with international arms control, disarmament and non-proliferation efforts to which they are party. It also enhances global efforts to protect and defend against CBRN threats and improve crisis management and recovery if WMD are employed against the Alliance or its interests.

Of particular importance is NATO's outreach to and cooperation with the United Nations (UN), the European Union (EU), and other regional organisations and multilateral initiatives that address WMD proliferation. Continued cooperation with regional organisations such as the Organization for Security and Co-operation in Europe (OSCE) can contribute to efforts to encourage member states to comply with relevant international agreements.

On the practical side, NATO organises an annual non-proliferation conference involving a significant number of non-member countries from six continents. This event is unique among international institutions' activities in the non-proliferation field, as it provides a venue for informal discussions among senior national officials on all types of WMD threats, as well as potential political and diplomatic responses. The conference has been hosted by both Allies and partners since it first took place at the NATO Defence College in Rome in 2004, followed by events in Sofia, Vilnius, Berlin, Warsaw, Prague, Bergen, Budapest, Split, Interlaken, Doha, Ljubljana and Helsinki.

The Alliance also participates in relevant conferences organised by other international institutions, including the UN Office for Disarmament Affairs, the EU, the Organization for the Prohibition of Chemical Weapons, the Biological and Toxin Weapons Convention, the OSCE, and others.

Many of NATO's activities under the SPS Programme focus on the civilian side of nuclear, chemical and biological technology. Scientists from NATO and partner countries are cooperating in research that impacts on these areas. Some examples include the decommissioning and disposal of WMD or their components, the safe handling of materials, techniques for arms control implementation, and the detection of CBRN agents.

The decision-making bodies:

The North Atlantic Council, NATO's principal political decision-making body, has overall authority on Alliance policy and activity in countering WMD proliferation. The

Council is supported by a number of NATO committees and groups, which provide strategic assessments and policy advice and recommendations.

The Committee on Proliferation is the senior advisory body for discussion of the Alliance's political and defence efforts against WMD proliferation. It brings together senior national officials responsible for political and security issues related to non-proliferation with experts on military capabilities needed to discourage WMD proliferation, to deter threats and the use of such weapons and to protect NATO populations, forces and territories. The Committee on Proliferation is chaired by NATO's International Staff when discussing political-military aspects of proliferation, and by national co-chairs when discussing defence-related issues.

Evolution:

The use or threatened use of WMD significantly influenced the security environment of the 20th century and will also impact international security in the foreseeable future. Strides in modern technology and scientific discoveries have opened the door to even more destructive weapons.

During the Cold War, the use of nuclear weapons was prevented by the prospect of mutually assured destruction. The nuclear arms race slowed in the early 1970s following the negotiation of the first arms control treaties.

The improved security environment of the 1990s enabled nuclear weapon states to dramatically reduce their nuclear stockpiles. However, the proliferation of knowledge and technology has enabled other countries to build their own nuclear weapons, extending the overall risks to new parts of the world.

At the Washington Summit in 1999, Allied leaders launched a Weapons of Mass Destruction Initiative to address the risks posed by the proliferation of these weapons and their means of delivery. The initiative was designed to promote understanding of WMD issues, develop ways of responding to them, improve intelligence and information-sharing, enhance existing Allied military readiness to operate in a WMD environment and counter threats posed by these weapons. Consequently, the WMD Non-Proliferation Center was established in 2000.

At the 2002 Prague Summit, Allies launched a modernisation process which aimed to ensure that the Alliance is able to effectively meet the new challenges of the 21st century. This included the creation of the NATO Response Force, the streamlining of the Alliance command structure and a series of measures to protect NATO populations, forces and territories from chemical, biological, radiological and nuclear (CBRN) threats.

In 2003, NATO created the Multinational CBRN Defence Battalion and Joint Assessment Team, which have been part of the Combined Joint CBRN Defence Task Force since 2007.

At the Riga Summit in 2006, Allied leaders endorsed a Comprehensive Political Guidance (CPG) that provides an analysis of the future security environment and a fundamental vision for NATO's ongoing transformation. It explicitly mentions the proliferation of WMD and their means of delivery as major security threats, which are particularly dangerous when combined with the threats of terrorism or failed states.

In July 2007, NATO activated a Joint CBRN Defence Center of Excellence in Vyškov, Czech Republic.

In April 2009, NATO Heads of State and Government endorsed NATO's "Comprehensive Strategic-Level Policy for Preventing the Proliferation of WMD and Defending against CBRN Threats". On 31 August 2009, the North Atlantic Council decided to make this document public.

At the November 2010 Lisbon Summit, Allied leaders adopted a new Strategic Concept. They also agreed at Lisbon to establish a dedicated committee providing advice on arms control, disarmament and non-proliferation. This committee started work in March 2011.

In May 2012 at the Chicago Summit, NATO leaders approved and made public the results of the Deterrence and Defence Posture Review. This document reiterates NATO's commitment "to maintaining an appropriate mix of nuclear, conventional and missile defence capabilities for deterrence and defence to fulfil its commitments as set out in the Strategic Concept". The Summit also reaffirmed that arms control, disarmament and non-proliferation play an important role in the achievement of the Alliance's security objectives and therefore Allies will continue to support these efforts.

Allied Heads of State and Government further emphasised that "proliferation threatens our shared vision of creating the conditions necessary for a world without nuclear weapons in accordance with the goals of the Nuclear Non-Proliferation Treaty (NPT)".

At the 2016 Warsaw Summit, Allies stated that they will ensure that NATO continues to be both strategically and operationally prepared with policies, plans and capabilities to counter a wide range of state and non-state CBRN threats.

In short:

1.- NATO Allies seek to prevent the proliferation of WMD through an active political agenda of arms control, disarmament and non-proliferation.

2.- The Arms Control, Disarmament, and WMD Non-proliferation Center (ACDC) at NATO Headquarters, strengthens dialogue among Allies, assesses risks to Allied populations, forces and territories, and supports chemical, biological, radiological or nuclear defence efforts.

3.- NATO is strengthening its capabilities to defend against chemical, biological, radiological or nuclear (CBRN) attacks, including terrorism and warfare.

4.- NATO conducts training and exercises designed to test interoperability and prepare forces to operate in a CBRN environment.


**NATO's response to hybrid threats**

Hybrid methods of warfare, such as propaganda, deception, sabotage and other non-military tactics have long been used to destabilise adversaries. What is new about attacks seen in recent years is their speed, scale and intensity, facilitated by rapid technological change and global interconnectivity. NATO has a strategy on its role in countering hybrid warfare and stands ready to defend the Alliance and all Allies against any threat, whether conventional or hybrid.

Hybrid threats combine military and non-military as well as covert and overt means, including disinformation, cyber attacks, economic pressure, deployment of irregular armed groups and use of regular forces. Hybrid methods are used to blur the lines between war and peace, and attempt to sow doubt in the minds of target populations.

The speed, scale and intensity of hybrid threats have increased in recent years. Being prepared to prevent, counter and respond to hybrid attacks, whether by state or non-state actors, is a top priority for NATO.

Since 2015, NATO has had a strategy on its role in countering hybrid warfare. NATO will ensure that the Alliance and Allies are sufficiently prepared to counter hybrid attacks in whatever form they may materialise. It will deter hybrid attacks on the Alliance and, if necessary, will defend Allies concerned.

To be prepared, NATO continuously gathers, shares and assesses information in order to detect and attribute any ongoing hybrid activity. The Joint Intelligence and Security Division at NATO Headquarters improves the Alliance's understanding and analysis of hybrid threats. The hybrid analysis branch provides decision-makers with improved awareness on possible hybrid threats.

The Alliance supports Allies 'efforts to identify national vulnerabilities and strengthen their own resilience, if requested. NATO also serves as a hub for expertise, providing support to Allies in areas such as civil preparedness and chemical, biological, radiological and nuclear (CBRN) incident response; critical infrastructure protection; strategic communications; protection of civilians; cyber defence; energy security; and counter-terrorism.

Training, exercises and education also play a significant role in preparing to counter hybrid threats. This includes exercising of decision-making processes and joint military and non-military responses in cooperation with other actors.

To deter hybrid threats, NATO is resolved to act promptly, whenever and wherever necessary. It continues to increase the readiness and preparedness of its forces, and has strengthened its decision-making process and its command structure as part of its deterrence and defence posture. This sends a strong signal that the Alliance is improving both its political and military responsiveness and its ability to deploy appropriate forces to the right place at the right time.

If deterrence should fail, NATO stands ready to defend any Ally against any threat. To this end, NATO forces have to be able to react in a quick and agile way, whenever and wherever needed.

NATO continues to strengthen its cooperation and coordination with such partners as Finland, Sweden, Ukraine and the European Union (EU) to counter hybrid threats and enhance resilience. As part of their increasingly closer cooperation, NATO and the EU have stepped up their cooperation on dealing with hybrid threats, with a special focus on countering cyber attacks.

In addition, Centers of Excellence work alongside and contribute knowledge and expertise to the Alliance. They are international research centers that are nationally or multi-nationally funded and staffed.

The European Center of Excellence for Countering Hybrid Threats located in Helsinki, Finland serves as a hub of expertise, assisting participating countries in improving their civil-military capabilities, resilience and preparedness to counter hybrid threats. It was inaugurated in October 2017 by NATO Secretary General Jens Stoltenberg, together with European Union High Representative for Foreign Affairs and Security Policy/Vice-President of the European Commission Federica Mogherini. The Center is an initiative of the Government of Finland, supported by 23 other nations, as well as NATO and the EU.

Other Centers of Excellence contribute to NATO's efforts to counter hybrid threats, including the Strategic Communications Center of Excellence in Riga, Latvia; the Cooperative Cyber Defence Center of Excellence in Tallinn, Estonia; and the Energy Security Center of Excellence in Vilnius, Lithuania.

## THE ALLIANCE'S ROLE IN DEVELOPING NEW CAPABILITIES

In this section we have to highlight that NATO is putting into place a series of measures to help improve the military capabilities of its member countries. Aimed at ensuring that the Alliance can fulfil its present and future operational commitments and fight new threats such as terrorism and the spread of weapons of mass destruction, these efforts build on a comprehensive array of measures taken since the end of the Cold War to adapt the Alliance to new challenges.

This is particularly important as NATO takes on new missions in faraway places like Afghanistan, which require forces that reach farther, faster, can stay in the field longer and can still undertake the most demanding operations if need be. Furthermore, these forces must be properly equipped and protected for the more dangerous missions they undertake.

Thus, in order to achieve these new objectives, the Alliance introduced three key initiatives that are the main driving force behind the transformation of the entire organisation: the Prague Capabilities Commitment to improve capabilities in critical areas such as strategic lift and air-to-ground surveillance; the streamlining of the military command structure; and the creation of the NATO Response Force.

Moreover, plans for an Alliance ground surveillance capability have moved forward and will provide situational awareness before and during NATO operations. To address the proliferation of weapons of mass destruction and their means of delivery, the Alliance is considering the possibility of missile defence for its territory, forces and population centres. It has also created a Multinational Chemical, Biological, Radiological and Nuclear (CBRN) Defence Battalion to defend against these threats.


**The Prague Capabilities Commitment**

We will emphasise that the Prague Capabilities Commitment (PCC) succeeded the Defence Capabilities Initiative (DCI) launched at the 1999 Washington Summit, which was designed to bring about improvements in the capabilities needed to ensure the effectiveness of future multinational operations across the full spectrum of Alliance missions, with a special focus on improving interoperability.

While DCI contributed to improvements in Alliance capabilities in quite a number of important areas, it was couched in terms of general commitments by member countries as a whole and did not require them to report individually on progress achieved. The 11 September 2001 attacks on the United States increased both the urgency and the importance of more focused capability improvements.

So, NATO made firm political commitments to improve capabilities in 400 specific areas which are fundamentally important to the efficient conduct of all Alliance missions. These cover the following eight fields:

• chemical, biological, radiological and nuclear (CBRN) defence;

• intelligence, surveillance and target acquisition;

• air-to-ground surveillance;

• command, control and communications;

• combat effectiveness, including precision-guided munitions and suppression of enemy air
defence;

• strategic airlift and sealift;

• air-to-air refuelling;

• deployable combat support and combat service support units.

Besides, defence ministers also decided that this new initiative should be based on firm country-specific commitments undertaken on the basis of national decisions and should incorporate target dates for the correction of shortfalls. In addition, such commitments should further increase multinational cooperation and role-sharing and should be realistic and achievable in economic terms, while representing a challenge to member countries. Moreover, they should achieve mutual reinforcement and full transparency with the related activities of the European Capability Action Plan initiated by the European Union.

Although the focus of the new initiative was sharper and involved individual commitments by member countries to specific capability improvements, to be contributed individually or together with other Allies, it concentrated on realistic and attainable objectives. The aim was clear: to deliver the urgently needed capability improvements to enable the Alliance to carry out all its missions, wherever they might occur.

Therefore, the PCC represents an important effort to ensure that Alliance forces have the means necessary to conduct operations swiftly and effectively for as long as necessary. It is by this agreement that heads of state and government at the Prague Summit  committed themselves to substantive capability improvements. While the principal responsibility for doing this lies with the governments of the member countries themselves, collectively the Alliance has put in place measures to track and monitor progress and take action to resolve any problems that arise. The PCC also seeks to identify ways of ensuring the mutual reinforcement of NATO's efforts and those of the European Union. Success in implementing these challenging but realistic and achievable goals is central to the fulfilment of the wider agenda laid down in Prague.

NATO command arrangements:

Likewise, a further central focus of the transformation process has been the streamlining of NATO's command arrangements. The function of the command structure is to plan and execute operations, to promote the modernisation and interoperability of Allied forces and to enhance the transatlantic link, which is at the heart of intra-Alliance cooperation. Changes to the command structure reflect these imperatives and assign a particularly important, continuous development role to Allied Command Transformation. This new command incorporates a NATO Joint Warfare Centre, a Joint Force Training Centre and a Joint Analysis and Lessons Learned Centre. Not part of the structure but linked to it are national and multinational Centres of Excellence that provide improved opportunities for training, interoperability, testing and developing military doctrines and assessing new concepts.

NATO Response Force:

According to official literature the establishment of the NATO Response Force (NRF) is an integral part of the transformation of NATO's military capabilities, complementing the Prague Capabilities Commitment and the new command structure.

The NRF was created with unprecedented speed. The significance of this achievement lies not only in the development of the force itself, but in the fact that its establishment affects other areas of capability improvement and acts as a catalyst for the sustained transformation and development of NATO forces as a whole. The NRF is one of the most important outcomes of the Prague Summit.

The NRF is a joint force of land, sea and air elements that can be tailored to individual missions and deployed rapidly wherever the North Atlantic Council requires. It is designed as a force that comprises technologically advanced, flexible, deployable, interoperable and sustainable elements, ready to deploy its leading elements within five days and able to sustain itself without further support for thirty days. It is not a permanent or standing force but one composed of units assigned by member countries in rotation, for set periods, and trained and certified together.

This force aims to prevent conflict or the threat of conflict from escalating into a wider dispute that threatens security and stability. It is capable of undertaking appropriate missions on its own or serving as part of a larger force contributing to the full range of Alliance military operations.

It could therefore be deployed in a number of different ways, for example as a show of force and demonstration of Alliance solidarity in the face of aggression; as a key element of a collective defence operation undertaken, in accordance with Article 5 of the North Atlantic Treaty, as a crisis management, peace support or stabilisation force for operations outside the framework of Article 5, or as an advance force for a larger-scale military operation, pending the deployment of other resources.

The prototype NRF comprising some 9500 troops was activated in October 2003. In the same timeframe, at their informal meeting in Colorado Springs, USA, NATO defence ministers participated in a study seminar designed to focus attention on the conceptual role of the NRF and the decision-making process relating to its potential deployment. At a ceremony in Brunssum, the Netherlands, on 15 October 2003, the new force was presented with its colours by General James Jones, Supreme Allied Commander Europe and the strategic commander of the newly established Allied Command Operations.

In an early demonstration of the force's initial capabilities, elements of the force including components from eleven NATO countries participated in a mock crisis response operation in Turkey in November 2003. The crisis involved a fictional threat to UN staff and to civilians located in a country outside the Euro-Atlantic area from terrorist activity and hostile forces. It called for an embargo on movements of forces and weapons, counterterrorist operations and a visible demonstration of Alliance solidarity, political determination and military capabilities.

Exercises and trials have continued to promote the development of the force on its planned schedule. In October 2004, during the informal meeting of NATO defence ministers in Poiana Brasov, Romania, the NATO Secretary General announced that the force had achieved its initial operational capability.

In September 2005, naval and air elements of the NRF were made available to the United States, following an official request for support after the extensive damage inflicted by Hurricane Katrina. The NRF was also used to provide humanitarian aid to Pakistan following the devastating earthquake of October 2005.

The force will be fully operational by October 2006, with a planned land element of about 21000 troops, and air and maritime elements of roughly the same size.

In short:

- The NATO Response Force (NRF) is a highly ready and technologically advanced, multinational force made up of land, air, maritime and Special Operations Forces (SOF) components that the Alliance can deploy quickly, wherever needed. In addition to its operational role, the NRF can be used for greater cooperation in education and training, increased exercises and better use of technology.

- Launched in 2002, the NRF consists of a highly capable joint multinational force able to react in a very short time to the full range of security challenges from crisis management to collective defence.

- NATO Allies decided to enhance the NRF in 2014 by creating a "spearhead force" within it, known as the Very High Readiness Joint Task Force (VJTF).

- This enhanced NRF is one of the measures of the Readiness Action Plan (RAP), which aims to respond to the changes in the security environment and strengthen the Alliance's collective defence.

- Overall command of the NRF belongs to the Supreme Allied Commander Europe (SACEUR).

- The decision to deploy the NRF is made by the North Atlantic Council, NATO's highest political decision-making body.

Alliance ground surveillance:

Significant progress in initiating improvements to capabilities in certain specific fields was made prior to the Prague Summit. In autumn 2002, plans for an Alliance ground surveillance (AGS) capability, a key element of NATO transformation, took a positive turn with the announcement of decisions on the cooperative development of a radar sensor designed to meet both the needs of the Alliance-owned AGS system and the national requirements of the countries participating in the development programme.

NATO is procuring an AGS system that will give Alliance commanders a picture of the situation on the ground in mission areas. It will consist of a mix of manned and unmanned radar platforms that can look down on the ground and relay data to commanders, providing them with "eyes in the sky" over a specific area.

The AGS system will be produced by the Transatlantic Industrial Partnership for Surveillance (TIPS), a consortium of over 80 companies, including the European Aeronautic Defence and Space Company (EADS), Galileo Avionica, General Dynamics Canada, Indra, Northrop Grumman and Thales. The system is scheduled to achieve an initial operational capability in 2010 and full capability in 2012.

Just as NATO's airborne warning and control system (AWACS) radar aircraft oversees airspace, AGS will be able to look at what is happening on the ground. AGS will provide situational awareness before and during NATO operations.

This is an essential capability for modern military operations and will be a key tool for the NATO Response Force (NRF).

To briefly summarise the meaning of what has come to be called NATO'S capabilities we will say that:

I.   NATO's modern defence posture is based on an effective combination of two key pillars: cutting-edge weapons systems and platforms, and forces trained to work together seamlessly.

II.  As such, investing in the right capabilities is an essential part of investing in defence. NATO plays an important role in assessing what capabilities the Alliance needs; setting goals for national or collective development of capabilities; and facilitating national, multinational and collective capability development and innovation.

III.    The Strategic Concept identifies collective defence, crisis management and cooperative security as NATO's core tasks. Deterrence based on an appropriate mix of nuclear, conventional and ballistic missile defence capabilities, remains a core element of NATO's overall strategy.

IV.    Allies have agreed to develop and maintain the full range of capabilities necessary to deter and defend against potential adversaries, using multinational approaches and innovative solutions where appropriate.

V.    The NATO Defence Planning Process (NDPP) is the primary means to identify and prioritise the capabilities required for full-spectrum operations, and to promote their development and delivery.

VI.    Developing and procuring capabilities through multinational cooperation helps generate economies of scale, reduces costs, and delivers interoperability by design. NATO actively supports Allies in the identification, launch and implementation of multinational cooperation.

VII.    To acquire vital capabilities, the Alliance must work closely with industry; build a stronger defence industry among Allies; foster greater industrial and technological cooperation across the Atlantic and within Europe; and maintain a robust industrial base throughout Europe and North America.


## NATO's new capabilities

*Air policing: securing NATO airspace*

- NATO Air Policing is a peacetime mission, which aims to preserve the security of Alliance airspace. It is a collective task and involves the continuous presence (24 hours a day, 365 days a year) of fighter aircraft and crews, which are ready to react quickly to airspace violations.

- NATO Air Policing is a collective task and a purely defensive mission, which involves the 24/7 presence of fighter aircraft, which are ready to react quickly to airspace violations.

- NATO members assist those Allies who are without the necessary means to provide air policing of their own territory.

- The Supreme Allied Commander Europe (SACEUR) is responsible for the conduct of the NATO Air Policing mission.

- Preservation of the integrity of NATO airspace is one of the missions of NATO Integrated Air and Missile Defence.

- Air policing was intensified following the Russia-Ukraine crisis.

*Alliance Ground Surveillance (AGS)*

Although we have previously referred to this point, we want to emphasise once again that it is part of these new NATO capabilities.

- NATO is acquiring the Alliance Ground Surveillance (AGS) system that will give commanders a comprehensive picture of the situation on the ground. A group of 15 Allies is acquiring the AGS system comprised of five NATO RQ-4D remotely piloted aircraft and the associated European-sourced ground command and control stations. NATO will then operate and maintain them on behalf of all NATO Allies. The AGS NATO RQ-4D aircraft is based on the US Air Force Block 40 Global Hawk. It has been uniquely adapted to NATO requirements to provide a state-of-the-art Intelligence, Surveillance and Reconnaissance (ISR) capability to NATO.

- The AGS system consists of air, ground and support segments, performing all-weather, persistent wide-area terrestrial and maritime surveillance in near real-time.

- AGS will provide in-theater situational awareness to commanders of deployed forces.

- AGS will be able to contribute to a range of missions such as protection of ground troops and civilian populations, border control and maritime safety, the fight against terrorism, crisis management and humanitarian assistance in natural disasters.

*AWACS: NATO's "eyes in the sky"*

- NATO operates a fleet of Boeing E-3A Airborne Warning & Control System (AWACS) aircraft, with their distinctive radar domes mounted on the fuselage, which provide the Alliance with air surveillance, command and control, battle space management and communications. NATO Air Base (NAB) Geilenkirchen, Germany, is home to 14 AWACS aircraft.

- NATO operates a fleet of Boeing E-3A Airborne Warning & Control System (AWACS) aircraft equipped with long-range radar and passive sensors capable of detecting air and surface contacts over large distances.

- The NATO Airborne Early Warning and Control Force (NAEW & C Force) is one of the few military assets that is actually owned and operated by NATO.

- It conducts a wide range of missions such as air policing, support to counter-terrorism, evacuation operations, embargo, initial entry and crisis response.

- Under normal circumstances, the aircraft operates for about eight hours, at 30,000 feet (9,150 meters) and covers a surveillance area of more than 120,000 square miles (310,798 square kilometers).

- The fleet is involved in the reassurance measures following the Russia-Ukraine crisis, and in the tailored assurance measures for Turkey against the background of the Syrian crisis.

- NATO AWACS aircraft are also providing surveillance and situational awareness to the Global Coalition to Defeat ISIS, thereby making the skies safer.
- AWACS surveillance aircraft played an important role in NATO operations such as in the United States after 9/11, in Libya and in Afghanistan. It also provided air support to secure NATO summits or international sporting events.

*Ballistic missile defence*

- Proliferation of ballistic missiles poses an increasing threat to NATO populations, territory and forces. Many countries have ballistic missiles or are trying to develop or acquire them. NATO ballistic missile defence (BMD) is part of the Alliance's response against the increasing threat and of its core task of collective defence.

- In 2010, Allies decided to develop a territorial BMD capability to pursue NATO's core task of collective defence.

- NATO has the responsibility to protect its European populations, territory and forces in light of the increasing proliferation of ballistic missiles and against threats emanating from outside the Euro-Atlantic area.

- NATO BMD is purely defensive; it is a long-term investment to address a long-term security threat.

- In July 2016, Allies declared Initial Operational Capability of NATO BMD, which offers a stronger capability to defend Alliance populations, territory and forces across southern NATO Europe against a potential ballistic missile attack.

- NATO BMD capability combines assets commonly funded by all Allies as well as voluntary contributions provided by individual Allies.

- Several Allies already offered their contributions or are undergoing development or acquisition of further BMD assets such as upgraded ships with BMD-capable radars, ground-based air and missile defence systems or advanced detection and alert capabilities.

*Boosting NATO's presence in the east and southeast*

- An important component of NATO's strengthened deterrence and defence posture is military presence in the eastern and south-eastern parts of Alliance territory. Allies implemented the 2016 Warsaw Summit decisions to establish NATO's forward presence in Estonia, Latvia, Lithuania and Poland and to develop a tailored forward presence in the Black Sea region.

- NATO has enhanced its forward presence in the eastern part of the Alliance, with four multinational battalion-size battlegroups in Estonia, Latvia, Lithuania and Poland, on a rotational basis.

- These battlegroups, led by the United Kingdom, Canada, Germany and the United States respectively, are robust, multinational, combat-ready forces.

- They demonstrate the strength of the transatlantic bond and make clear that an attack on one Ally would be considered an attack on the whole Alliance.

- It is part of the biggest reinforcement of Alliance collective defence in a generation.

- NATO has also a forward presence tailored to the southeast of Alliance territory and in the Black Sea region. Allies are contributing forces and capabilities on land, at sea and in the air.

- The land element in the southeast of the Alliance is built around a multinational brigade, under Multinational Division Southeast in Romania and is coordinating multinational training through a Combined Joint Enhanced Training Initiative.

- In the air, several Allies have reinforced Romania's and Bulgaria's efforts to protect NATO airspace.

*Centers of Excellence*

- Centers of Excellence (COEs) are international military organisations that train and educate leaders and specialists from NATO member and partner countries. They assist in doctrine development, identify lessons learned, improve interoperability and capabilities, and test and validate concepts through experimentation.

- They offer recognised expertise and experience that is of benefit to the Alliance, and support the transformation of NATO, while avoiding the duplication of assets, resources and capabilities already present within the Alliance.

- COEs cover a wide variety of areas such as civil-military operations, cyber defence, military medicine, energy security, naval mine warfare, defence against terrorism, cold weather operations, and counter-IED.

- Allied Command Transformation has overall responsibility for COEs and is in charge of the establishment, accreditation, preparation of candidates for approval, and periodic assessments of the centers.

- COEs are nationally or multi-nationally funded. NATO does not directly fund COEs nor are they part of the NATO Command Structure.

*Civil preparedness*

- The Alliance depends on civil and commercial resources and infrastructure, such as railways, ports, airfields and energy grids, to support the rapid and effective movement and sustainment of its military forces. Such assets are vulnerable to external attack and internal disruption.

- Civil preparedness means that basic government functions can continue during emergencies or disasters, in peacetime or in periods of crisis. It also means that the civilian sector in Allied nations would be ready to provide support to a NATO military operation.

- Under Article 3 of the North Atlantic Treaty, all Allies are committed to building national resilience, which is the combination of civil preparedness and military capacity.

- Allies agreed baseline resilience requirements in seven strategic sectors: continuity of government, energy, population movement, food and water resources, mass casualties, civil communications and transport systems.

- As part of its efforts to strengthen resilience, NATO has worked with Allies to enhance preparedness across the whole of government, including in the health sector, by providing general guidelines.

- To deter, counter or recover from threats or disruptions to the civil sector, effective action requires clear plans and response measures, defined well ahead of time and exercised regularly.
- That is why there is a need to complement military efforts to defend Alliance territory and populations with robust civil preparedness.

*Combined Joint Chemical, Biological, Radiological and Nuclear Defence Task Force*

Although we have previously referred to this point, we want to emphasise once again that it is part of these new NATO capabilities.

- NATO today faces a whole range of complex challenges and threats to its security. Current threats include the proliferation of weapons of mass destruction (WMD) and their delivery systems. Rapid advances in biological science and technology also continue to increase the bio-terrorism threat against NATO forces and populations. The Alliance needs to be prepared to prevent, protect and recover from WMD attacks or CBRN events.

- NATO's Combined Joint CBRN Defence Task Force consists of the CBRN Joint Assessment Team and the CBRN Defence Battalion.

- The CBRN Defence Battalion is a NATO body specifically trained and equipped to deal with CBRN incidents and / or attacks against NATO populations, territory or forces.

- The Battalion trains not only for armed conflict, but also for deployment in crises, where it supports civilian authorities, such as natural disasters and industrial accidents.

- It falls under the authority of the Supreme Allied Commander Europe.

*Connected Forces Initiative*

- The Connected Forces Initiative (CFI) aims to enhance the high level of interconnectedness and interoperability Allied forces have achieved on operations and with partners. CFI combines a comprehensive education, training, exercise and evaluation program with the use of cutting-edge technology to ensure that Allied forces remain prepared to engage cooperatively in the future.

- CFI is a key enabler in developing the goal of NATO Forces 2020: a coherent set of deployable, interoperable and sustainable forces equipped, trained, exercised, commanded and able to operate together and with partners in any environment.

- The Initiative is essential in ensuring that the Alliance remains well prepared to undertake the full range of its missions, as well as to address future challenges wherever they may arise.

- In light of the current security environment, it is also a means to deliver the training and exercise elements of the Alliance's Readiness Action Plan.

*Countering terrorism*

Although we have previously referred to this point, we want to emphasise once again that it is part of these new NATO capabilities.

- Terrorism in all its forms poses a direct threat to the security of the citizens of NATO countries, and to international stability and prosperity.

- It is a persistent global threat that knows no border, nationality or religion and is a challenge that the international community must tackle together.

- NATO will continue to fight this threat in all its forms and manifestations with determination and in full solidarity.

- NATO's work on counter-terrorism focuses on improving awareness of the threat, developing capabilities to prepare and respond, and enhancing engagement with partner countries and other international actors.

- NATO invoked its collective defence clause (Article 5) for the first and only time in response to the terrorist attacks of 11 September 2001 on the United States.

- NATO's Counter-Terrorism Policy Guidelines focus Alliance efforts on three main areas: awareness, capabilities and engagement.

- A comprehensive Action Plan is being implemented to enhance NATO's role in the international community's fight against terrorism.

- NATO has a Terrorism Intelligence Cell at NATO Headquarters and a Coordinator oversees NATO's efforts in the fight against terrorism.

- A regional Hub for the South, based at NATO's Joint Force Command in Naples helps the Alliance anticipate and respond to crises arising in its southern neighborhood.

- NATO is a member of the Global Coalition to Defeat ISIS and supports it through AWACS intelligence flights.

- NATO develops new capabilities and technologies to tackle the terrorist threat and to manage the consequences of a terrorist attack.

- NATO cooperates with partners and international organisations to leverage the full potential of each stakeholder engaged in the global counter-terrorism effort.

- Cyber threats to the security of the Alliance are becoming more frequent, complex, destructive and coercive. NATO will continue to adapt to the evolving cyber threat landscape. NATO and its Allies rely on strong and resilient cyber defences to fulfil the Alliance's core tasks of collective defence, crisis management and cooperative security. The Alliance needs to be prepared to defend its networks and operations against the growing sophistication of the cyber threats and attacks it faces.

- Cyber defence is part of NATO's core task of collective defence.

- NATO has affirmed that international law applies in cyberspace.

- NATO's main focus in cyber defence is to protect its own networks (including operations and missions) and enhance resilience across the Alliance.

- In July 2016, Allies reaffirmed NATO's defensive mandate and recognised cyberspace as a domain of operations in which NATO must defend itself as effectively as it does in the air, on land and at sea.

- Allies also made a Cyber Defence Pledge in July 2016 to enhance their cyber defences, as a matter of priority. Since then, all Allies have upgraded their cyber defences.

- NATO reinforces its capabilities for cyber education, training and exercises.

- Allies are committed to enhancing information-sharing and mutual assistance in preventing, mitigating and recovering from cyber attacks.

- NATO Cyber Rapid Reaction teams are on standby to assist Allies, 24 hours a day, if requested and approved.

- At the Brussels Summit in 2018, Allies agreed to set up a new Cyberspace Operations Center as part of NATO's strengthened Command Structure. They also agreed that NATO can draw on national cyber capabilities for its missions and operations.

- In February 2019, Allies endorsed a NATO guide that sets out a number of tools to further strengthen NATO's ability to respond to significant malicious cyber activities.

- NATO and the European Union (EU) are cooperating through a Technical Arrangement on Cyber Defence, which was signed in February 2016. In light of common challenges, NATO and the EU are strengthening their cooperation on cyber defence, notably in the areas of information exchange , training, research and exercises.

- NATO is intensifying its cooperation with industry through the NATO Industry Cyber Partnership.

- NATO recognises that its Allies stand to benefit from a norms-based, predictable and secure cyberspace.

*Education and training*

- In order to safeguard the freedom and security of its members, the Alliance must maintain the capabilities to prevent, detect, deter and defend against any threat of aggression. For this reason, NATO conducts education and training programs to increase cohesion, effectiveness and readiness of its multinational forces. Furthermore, NATO shares its expertise with partner countries in their education and training reform efforts.

- Since its inception in 1949, NATO has engaged in education and training activities, which have expanded geographically and institutionally over time.

- The creation of Allied Command Transformation (ACT) in 2002 is testimony of NATO's resolve to boost education and training. ACT is entirely dedicated to leading the transformation of NATO's military structure, forces, capabilities and doctrine, including through exercise and training design and management.

- ACT has a holistic approach to education and training: it provides unity of effort and helps identify gaps and avoid duplication, while ensuring greater effectiveness and efficiency through global programming. These efforts are complementary to national programmes.

- Supreme Allied Commander Europe (SACEUR), at the head of Allied Command Operations (ACO), provides strategic-level guidance and sets the priorities and requirements for NATO Education, Training, Exercises and Evaluation (ETEE).

- NATO's education and training programmes help to improve "interoperability" of multinational forces, i.e. their ability to work together.

- The Alliance is also committed to effective cooperation and coordination with partner countries and international organisations, such as the United Nations, the European Union and the African Union.

*Electronic warfare*

- Electronic warfare (EW) capabilities are a key factor in the protection of military forces and in monitoring compliance with international agreements.

- They are essential for the full spectrum of operations and other tasks undertaken by the Alliance.

- The purpose of EW is to deny the opponent the advantage of, and ensure friendly unimpeded access to the electromagnetic spectrum. EW can be applied from air, sea, land and space, and target communication and radar systems. It involves the use of the electromagnetic energy to provide improved understanding of the operational environment as well as to achieve specific effects on the modern battlefield.

- The need for military forces to have unimpeded access to and use of the electromagnetic environment creates challenges and opportunities for EW in support of military operations.

- The NATO Electronic Warfare Advisory Committee (NEWAC) is responsible for overseeing the development of NATO's EW policy, doctrine, and command and control concepts as well as monitoring EW support to NATO operations.

- It also assists in introducing NATO's EW concepts to partner countries within the framework of the Partnership for Peace program.

- The NEWAC is composed of representatives of each NATO country and of the Strategic Commands. Members are senior officials in national electronic warfare organisations.

- The Chairman and the Secretary of the committee are permanently assigned to the International Military Staff at NATO Headquarters, Brussels.

- There are a number of subordinate groups dealing with electronic warfare database support, training and doctrine.

- Exercises are important tools through which the Alliance tests and validates its concepts, procedures, systems and tactics. More broadly, they enable militaries and civilian organisations deployed in theaters of operation to test capabilities and practice working together efficiently in a demanding crisis situation.

- Exercises allow NATO to test and validate concepts, procedures, systems and tactics.

- They enable militaries and civilian organisations deployed on the ground to work together to identify "best practices" (what works) and "lessons learned" (what needs improving).

- Exercises also contribute to improved interoperability and defence reform.

- NATO has recently boosted its exercise program in light of the changed security environment.

- Exercises are planned in advance and vary in scope, duration and form - ranging between live exercises in the field to computer-assisted exercises that take place in a classroom.

- To foster and support interoperability, NATO exercises are as open as possible to all formal partner countries.

- The Alliance has been conducting exercises since 1951.


*Joint Intelligence, Surveillance and Reconnaissance*

- Joint Intelligence, Surveillance and Reconnaissance (JISR) is vital for all military operations. It provides decision-makers and action-takers with a better situational awareness of what is happening on the ground, in the air or at sea. This means that Allies work together to collect, analyse and share information to maximum effect. This makes Joint ISR a unique example of cooperation and burden-sharing across the Alliance.

- NATO is establishing a permanent JISR system providing information and intelligence to key decision-makers, helping them make well-informed, timely and accurate decisions.

- JISR brings together data and information gathered through projects such as NATO's Alliance Ground Surveillance (AGS) system or NATO AWACS surveillance aircraft as well as a wide variety of national JISR assets from the space, air, land and maritime domains.

- Both surveillance and reconnaissance includes visual observation (from soldiers on the ground) and electronic observation (for example from satellites, unmanned aircraft systems, ground sensors and maritime vessels), which are then analysed, turning information into intelligence.

- The Initial Operational Capability (IOC) for JISR, declared in February 2016, represents a significant achievement, enabling better connectivity between NATO and Allies' capabilities.

- IOC is only the first milestone for the JISR initiative. Further work is needed to sustain these achievements.

*Logistics*

- While the term "logistics" can encompass several different meanings, in essence it has to do with having the right thing, at the right place, at the right time. NATO defines logistics as the science of planning and carrying out the movement and maintenance of forces. It is of vital importance for any military operation and, without it, operations could not be carried out and sustained. Logistics can be seen as the bridge between deployed forces and the industrial base, which produces the material and weapons deployed forces need to accomplish their mission.

- The services and responsibilities of NATO logistics are subdivided into three domains: production logistics, in-service logistics, and consumer logistics.

- Multinational logistics is a component of collective logistics, which aims to achieve reduction in costs, harmonise life-cycle processes and increase efficiency in logistics support at all times.

- NATO logistics can also be understood through the core functions they fulfil which include but are not limited to: supply, maintenance, movement and transportation, petroleum support, infrastructure engineering, and medical support.

- One of the key logistics principles driving logistic support at NATO is that of collective responsibility which encourages nations and NATO to cooperatively share the provision and use of logistic capabilities and resources.

- The Logistics Committee is the principal committee that supports the North Atlantic Council and the Military Committee as the overarching coordinating authority across the whole spectrum of logistics functions within NATO.

- In the wake of the Russia-Ukraine conflict since 2014, the NATO Logistics Vision and Objectives (V&O) was revised in accordance with developments from the 2010 Strategic Concept, Political Guidance 2015, and the Readiness Action Plan.

*Military medical support*

- The primary responsibility of military medical services is preserving and maintaining health and fighting strength of the military. At the same time, through civil-military cooperation, the military medical services strengthen and facilitate civilian efforts in dealing with medical issues such as the coronavirus pandemic. With the outbreak of COVID-19, fighting the pandemic became one of their main priorities. Their role in coordinating national efforts within the military in NATO member and partner countries and their support to civilian efforts are fundamental contributions to managing the crisis. In addition to these roles, medical support is one of the key planning domains for operations.

- The military health support system aims to preserve and restore the health of NATO personnel and consequently to contribute to preserving the operational capacity of NATO member and partner forces at home and in deployment.

- Civil-military cooperation is vital at all times. With the outbreak of the coronavirus pandemic, the military medical services support civilian authorities, which are in the lead in responding to the crisis.

- The Committee of the Chiefs of Military Medical Services in NATO (COMEDS) is the senior body for military medical advice within NATO.

- COMEDS (in close cooperation with the Medical Advisors of the NATO Command Structure) is the central point for the development and coordination of military medical matters and for providing medical advice to the Military Committee.

- For COVID-19, it helped to coordinate military medical aspects of the pandemic among members and partner countries in order to identify issues that require harmonisation, immediate attention, decision or action.

- It also develops new concepts of medical support for operations, with emphasis on multinational health care, modularity of medical treatment facilities and partnerships.

- In addition to COMEDS 'contribution, the NATO Center of Excellence for Military Medicine in Hungary plays a significant role by providing support through medical experts, training and advice, as well as through its coordination role.

- Recently, a new medical entity, the so-called Multinational Medical Coordination Center/European Medical Command in Germany, was created.

*Meteorology and oceanography*

- Today, the Alliance is often operating, or monitoring conditions that affect its strategic interests, beyond the borders of its member nations. It therefore needs to have the most accurate, timely and relevant information (both current and forecast) describing the meteorological and oceanographic (METOC) aspects of these environments. For example, comprehensive weather and flood forecasting and oceanographic features such as wave heights, temperature, salinity, surf and tidal movements, or even the presence of marine life, can seriously affect military activities.

- NATO cooperation in METOC support for its forces aims to ensure that Allies get the information they need through efficient and effective use of national and NATO assets.

- This information helps allied forces exploit the best window of opportunity to plan, execute, support and sustain military operations. Furthermore, it helps them optimise the use of sensors, weapons, targeting, logistics, equipment and personnel.

- To advise the Military Committee, a METOC working group was recently formed from two separate meteorology and oceanography groups.

- The NATO Meteorological and Oceanographic Military Committee Working Group [MCWG (METOC)] advises the Military Committee on METOC issues.

- It also acts as a standardisation authority by supervising two subordinate panels on military meteorology and military oceanography.

- MCWG (METOC), which comprises delegates from each allied country, meets annually to address military METOC policy, procedures and standardisation agreements between NATO and partner countries. It relies to a large extent on the resources of NATO members, most of which have dedicated civil and/or military METOC organisations.

- The group supports NATO and national members in developing effective plans, procedures and techniques for providing METOC support to NATO forces and ensuring data is collected and shared.

- In a more general sense, it encourages research and development as well as liaison, mutual support and interoperability among national and NATO command METOC capabilities that support allied forces.

- NATO created the MCWG (METOC) by merging the former Military Oceanography Group and the Military Committee Meteorology Group in 2011.

- NATO member countries are expected to provide the bulk of METOC information and resources.

- At the same time, national delegates are able to steer policy, when needed, through the MCWG (METOC) and act as the approval authority for standardisation.

- Among other tasks, nations are expected to:

    - contribute to a network of data collection sites and platforms,

    - provide METOC analysis and forecasts, and

- provide military METOC support products and services, such as tactical decision aids (TDAs) and acoustic predictions.

*Multinational capability cooperation*

- To carry out its missions and tasks, NATO needs Allies to invest in interoperable, cutting-edge and cost-effective equipment. To that end, NATO plays an important role in helping nations decide how and where to invest in their defence.

- The Alliance also supports Allies in identifying and developing multinational cooperative projects to deliver the key defence capabilities needed for Alliance security.

- NATO is helping Allies to identify, initiate and advance opportunities for multinational capability cooperation in key areas such as air-to-air refueling, ammunition, maritime unmanned systems, command and control, and training.

- The aim is to drive down costs through economies of scale while improving operational values through increased commonality of equipment, training, doctrine and procedures.

- NATO Allies and partner nations have initiated several High Visibility Projects (HVPs), which are being developed.

- NATO works with the European Union to avoid duplication and ensure complementarity of efforts.

*NATO Air Command and Control System (ACCS)*

- The NATO Air Command and Control System (ACCS) program will provide the Alliance with a single, integrated air command and control system to manage NATO air operations in and out of the Euro-Atlantic area.

- NATO ACCS will replace a wide variety of NATO and national air systems currently fielded across the Alliance.

- It will provide a unified air command and control system, enabling NATO and its members to manage all types of air operations both over NATO European territory as well as when deployed out of area.

- Once fully deployed, NATO ACCS will cover 10 million square kilometers of airspace and interconnect over 20 military aircraft control centers.

*NATO Integrated Air and Missile Defence*

- NATO Integrated Air and Missile Defence (NATO IAMD) is an essential, continuous mission in peacetime, crisis and conflict, safeguarding and protecting Alliance territory, populations and forces against air and missile threat and attack. It contributes to deterrence and to indivisible security and freedom of action of the Alliance.

- NATO IAMD is the defensive part of the Alliance's Joint Air Power, which aims to ensure the stability and security of NATO airspace by coordinating, controlling and exploiting the air domain.

- It incorporates all measures to deter and defend against any air and missile threat or to nullify or reduce the effectiveness of hostile air action.

- NATO IAMD can address threats from the air, on land or at sea, which may include chemical, biological, radiological and nuclear, as well as electromagnetic and cyber threats.

- NATO IAMD provides a highly responsive, robust, time-critical and persistent capability in order to achieve a desired level of control of the air, where the Alliance is able to conduct the full range of its missions.

- NATO IAMD is implemented through the NATO Integrated Air and Missile Defence System (NATINAMDS), a network of interconnected national and NATO systems comprised of sensors, command and control facilities and weapons systems.

- NATINAMDS comes under the authority of NATO's Supreme Allied Commander Europe.

- NATO has a pipeline system designed to ensure that its requirements for petroleum products and their distribution can be met at all times.

- The NATO Pipeline System (NPS) was set up during the Cold War to supply NATO forces with fuel and it continues to satisfy fuel requirements with the flexibility that today's security environment requires.

- The NPS consists of ten distinct storage and distribution systems for fuels and lubricants.
- In total, it is approximately 10,000 kilometers long, runs through 12 NATO countries and has a storage capacity of 4.1 million cubic meters.

- The NPS links together storage depots, military air bases, civil airports, pumping stations, truck and rail loading stations, refineries and entry / discharge points.

- Bulk distribution is carried out using facilities from the common-funded NATO Security Investment Program.

- The networks are controlled by national organisations, with the exception of the Central Europe Pipeline System (CEPS), which is a multinational system managed by the CEPS Program Office under the aegis of the NATO Support and Procurement Agency.

Although we have previously referred to this point, we want to emphasise once again that it is part of these new NATO capabilities.

- The NATO Response Force (NRF) is a highly ready and technologically advanced, multinational force made up of land, air, maritime and Special Operations Forces (SOF) components that the Alliance can deploy quickly, wherever needed. In addition to its operational role, the NRF can be used for greater cooperation in education and training, increased exercises and better use of technology.

- Launched in 2002, the NRF consists of a highly capable joint multinational force able to react in a very short time to the full range of security challenges from crisis management to collective defence.

- NATO Allies decided to enhance the NRF in 2014 by creating a "spearhead force" within it, known as the Very High Readiness Joint Task Force (VJTF).

- This enhanced NRF is one of the measures of the Readiness Action Plan (RAP), which aims to respond to the changes in the security environment and strengthen the Alliance's collective defence.

- Overall command of the NRF belongs to the Supreme Allied Commander Europe (SACEUR).

- The decision to deploy the NRF is made by the North Atlantic Council, NATO's highest political decision-making body.

*NATO's approach to space*

- Space is a dynamic and rapidly evolving area, which is essential to the Alliance's deterrence and defence. In 2019, Allies adopted NATO's Space Policy and recognised space as a new operational domain, alongside air, land, sea and cyberspace. It will guide NATO's approach to space and ensure the right support to the Alliance's operations and missions in such areas as communications, navigation and intelligence. Through the use of satellites, Allies and NATO can respond to crises with greater speed, effectiveness and precision.

- Space is increasingly important to the Alliance's and Allies' security and prosperity.

- Space capabilities bring benefits in multiple areas from weather monitoring, environment and agriculture, to transport, science, communications and banking.

- The information gathered and delivered through space-based satellites is critical for NATO activities, operations and missions, including collective defence, crisis response, counter-terrorism and disaster relief.

- In 2019, Allies adopted a new Space Policy and declared space an operational domain.

- NATO remains a key forum for Allies to share information and coordinate activities on various space-related issues.

- NATO's approach to space will remain fully in line with international law.

*Rapid Air Mobility*

- NATO's Rapid Air Mobility (RAM) initiative enables Allied military aircraft to swiftly deploy on short notice across Europe, when activated, during crises. Aircraft are assigned a unique NATO call sign making it easier to move across borders with essential equipment, supplies and personnel, as quickly as possible and in a coordinated manner. In March 2020, in response to the COVID-19 pandemic, an adapted version of the RAM process was activated to enable the timely transport of medical supplies.

- NATO's Rapid Air Mobility initiative is part of its wider military mobility effort to boost the readiness and deployment of Allied troops and assets.

- Allied aircraft receive priority handling by the European Organization for the Safety of Air Navigation (EUROCONTROL) and can quickly deploy on short notice across Europe during a crisis.

- RAM can be activated during crises in peacetime.

- The North Atlantic Council has overall political control over RAM, while the authority to activate and run it operationally belongs to SACEUR, one of the two Strategic Commanders.

- In March 2020, in response to the outbreak of the COVID-19 virus, the initiative was activated to ensure timely delivery of medical supplies to Allies in need.

*Rapid Deployable Corps*

- NATO's Rapid Deployable Corps are High Readiness Headquarters, which can be quickly dispatched to lead NATO troops on missions within or beyond the territory of NATO member states.

- NATO's Rapid Deployable Corps are High Readiness Headquarters, which can be quickly dispatched to lead NATO troops wherever necessary.

- The corps can be deployed for a wide range of missions: from disaster management, humanitarian assistance and peace support to counter-terrorism and high-intensity war fighting.

- There are currently nine NATO Rapid Deployable Corps, which are each capable of commanding up to 60,000 soldiers.

- The political authorisation of the North Atlantic Council (NAC), NATO's principal political decision-making body, is required to deploy the corps.

*Readiness Action Plan*

- The Readiness Action Plan (RAP) ensures that the Alliance is ready to respond swiftly and firmly to new security challenges from the east and the south. Begun at the 2014 Wales Summit, this is the most significant reinforcement of NATO's collective defence since the end of the Cold War. At Warsaw in 2016, Allied leaders welcomed its implementation and introduced new work on NATO's deterrence and defence posture.

- Due to the changed security environment on NATO's borders, the RAP includes 'assurance measures' for NATO member countries in Central and Eastern Europe to reassure their populations, reinforce their defence and deter potential aggression.
- Assurance measures comprise a series of land, sea and air activities in, on and around the eastern part of Alliance territory, which are reinforced by exercises focused on collective defence and crisis management.

- The RAP also includes 'adaptation measures' which are longer-term changes to NATO's forces and command structure so that the Alliance will be better able to react swiftly and decisively to sudden crises.

- Adaptation measures included tripling the size of the NATO Response Force (NRF), the establishment of a Very High Readiness Joint Task Force (VJTF) able to deploy at very short notice, and enhanced Standing Naval Forces.

- To facilitate readiness and the rapid deployment of forces, eight NATO Force Integration Units (NFIUs) - which are small headquarters - were established in Central and Eastern Europe.

- Headquarters for the Multinational Corps Northeast in Szczecin, Poland and the Multinational Division Southeast in Bucharest, Romania were also established. In addition, a standing joint logistics support group headquarters was set up.

- At the 2016 Warsaw Summit, Allies welcomed the implementation of the RAP and agreed to further strengthen the Alliance's deterrence and defence posture with an enhanced forward presence in the eastern and southeast part of Alliance territory and a framework for NATO's adaptation in response to growing challenges and threats emanating from the south.

- NATO's Enhanced Forward Presence is made up of four battlegroups in Estonia, Latvia, Lithuania and Poland. These battlegroups are multinational and combat-ready, demonstrating the strength of the transatlantic bond.

- Each NATO member country needs to be resilient to resist and recover from a major shock such as a natural disaster, failure of critical infrastructure, or a hybrid or armed attack. Resilience is a society's ability to resist and recover easily and quickly from such shocks and combines both civilian preparedness and military capacity. Robust resilience through civil preparedness in Allied countries is essential to NATO's collective security and defence.

- The principle of resilience is firmly anchored in Article 3 of the Alliance's founding treaty: "In order more effectively to achieve the objectives of this Treaty, the Parties, separately and jointly, by means of continuous and effective self-help and mutual aid, will maintain and develop their individual and collective capacity to resist armed attack".

- Resilience is first and foremost a national responsibility and each member country needs to be sufficiently robust and adaptable to support the entire spectrum of crises envisaged by the Alliance. In this context, Article 3 complements the collective defence clause set out in Article 5, which stipulates that an attack against one Ally is an attack against all. Allies need to give NATO the means to fulfil its core tasks and, in particular, that of collective defence.

- The individual commitment of each and every Ally to maintain and strengthen its resilience reduces the vulnerability of NATO as a whole. Members can strengthen resilience through the development of home defence and niche skills such as cyber defence or medical support combining civilian, economic, commercial and military factors. When Allies are well prepared, they are less vulnerable or less likely to be attacked, making NATO as a whole stronger.

- Moreover, military forces, especially those deployed during crises and war, heavily dependent on the civilian and commercial sectors for transport, communications and even basic supplies such as food and water, to fulfil their missions. Military efforts to defend Alliance territory and populations therefore need to be complemented by robust civil preparedness. However, civil capabilities can be vulnerable to disruption and attack in both peacetime and during war. By reducing these vulnerabilities, NATO reduces the risk of a potential attack. A high level of resilience is therefore an essential aspect of credible deterrence and defence.

*Smart Defence*

- In these times of austerity, each euro, dollar or pound sterling counts. Smart Defence is a cooperative way of thinking about generating the modern defence capabilities that the Alliance needs for the future. In this renewed culture of cooperation, Allies are encouraged to work together to develop, acquire, operate and maintain military capabilities to undertake the Alliance's essential core tasks agreed in NATO's Strategic Concept. That means harmonising requirements, pooling and sharing capabilities, setting priorities and coordinating efforts better.

- Smart Defence is a cooperative way of generating modern defence capabilities that the Alliance needs, in a more cost-efficient, effective and coherent manner.

- Allies are encouraged to work together to develop, acquire, operate and maintain military capabilities to undertake the Alliance's core tasks.

- Projects cover a wide range of efforts addressing the most critical capability requirements such as precision-guided munitions, cyber defence, ballistic missile defence, and Joint Intelligence, Surveillance and Reconnaissance to name a few.

*Special Operations Forces*

- NATO Special Operations Forces (SOF) provide capabilities that complement those of NATO air, maritime and land forces and are relevant across the full range of military operations. These SOF capabilities are also applicable to the Alliance's core tasks of collective defence, crisis management and cooperative security. The NATO Special Operations Headquarters (NSHQ) is the primary point of development, coordination and direction for all NATO Special Operations activities.

- Located at Supreme Headquarters Allied Powers Europe (SHAPE) in Mons, Belgium and under the daily direct operational command of the Supreme Allied Commander Europe (SACEUR), the NSHQ focuses on ensuring Allied Joint1 SOF personnel possess a multinational foundation to allow them to operate as effectively, efficiently and coherently as possible in support of the Alliance's objectives from the strategic to the tactical level. Twenty-six NATO member countries and three partners (Austria, Finland and Sweden) are represented among 200 plus headquarters staff.

- The NSHQ is a unique hybrid organisation. It is involved in a very diverse set of activities such as NATO SOF policy, doctrine, capabilities, standards, training and education. On a daily basis the NSHQ is actively coordinating, advocating and advising reference SOF across NATO. These activities include areas such as SOF-specific intelligence, aviation, medical support and communications.

- The NSHQ also supports SOF involvement in NATO operations.

- The NSHQ provides an additional deployable NATO SOF command and control option to complement other existing mechanisms provided by NATO member countries for the NRF.

*Standardization*

- The ability to work together is more important than ever for the Alliance. States need to share a common set of standards, especially among military forces, to carry out multinational operations. By helping to achieve interoperability among NATO's forces, as well as with those of its partners, standardisation allows for more efficient use of resources and thus enhances the Alliance's operational effectiveness.

- To work together effectively and efficiently, NATO forces as well as partner forces need to share common set of standards.

- Standardisation allows for more efficient use of resources and thus enhances the effectiveness of the Alliance's defence capabilities.

- A Standardization Agreement (STANAG) is a NATO standardisation document that specifies the agreement of member nations to implement a standard.

- The Committee for Standardization (CS) is the senior NATO committee for Alliance standardisation, composed primarily of representatives from all NATO countries. Operating under the authority of the North Atlantic Council (NAC), it issues policy and guidance for all NATO standardisation activities. Its mission is to exert domain governance for standardisation policy and management within the Alliance to contribute to Allies 'development of interoperable and cost-effective military forces and capabilities.

- The NATO Standardization Office (NSO) initiates, coordinates, supports and administers NATO standardisation activities conducted under the authority of the Committee for Standardization (CS). It also assists NATO's Military Committee in developing military operational standardisation. Its mission is to foster NATO standardisation with the goal of enhancing the operational effectiveness of Alliance military forces.

- The NATO Standardization Staff Group (NSSG) assists the Director of the NSO. It is a staff-level forum which facilitates coherence of NATO standardisation activities and procedures across NATO bodies, especially the standardisation tasking authorities.

*Strategic airlift*

- NATO member countries are pooling their resources to charter special aircraft that give the Alliance the capability to transport troops, equipment and supplies across the globe. Robust strategic airlift capabilities are vital to ensure that NATO countries are able to deploy their forces and equipment rapidly to wherever they are needed.

- By pooling resources, NATO countries make significant financial savings and have the potential of acquiring assets collectively that would be prohibitively expensive to purchase as individual countries.

- There are currently two initiatives aimed at providing the Alliance with strategic airlift capabilities: the Strategic Airlift International

Solution (SALIS) initiative, and the Strategic Airlift
Capability (SAC).

*Strategic sealift*

- NATO member countries have pooled their
  resources to assure access to special ships, giving the
  Alliance the capability to rapidly transport forces
  and equipment by sea.

- This multinational consortium finances the charter of
  up to 11 special "roll-on/roll-off" ships (commonly,
  Ro/Ro; so called because equipment can be driven
  onto and off of the ships via special doors and ramps
  into the hold) . The consortium includes Belgium,
  Canada, Denmark, Germany, Hungary, Lithuania,
  the Netherlands, Norway, Portugal, Slovenia and the
  United Kingdom.

*Weapons of mass destruction*

Although we have previously referred to this point, we want
to emphasise once again that it is part of these new NATO
capabilities

- The proliferation of nuclear weapons and other
  weapons of mass destruction (WMD), and their
  delivery systems, could have incalculable
  consequences for national, regional and global
  security. During the next decade, proliferation will
  remain most acute in some of the world's most
  volatile regions. The potential effects of WMD

proliferation on NATO Allies are one of the greatest threats NATO faces.

- NATO Allies seek to prevent the proliferation of WMD through an active political agenda of arms control, disarmament and non-proliferation.

- The Arms Control, Disarmament, and WMD Non-proliferation Center (ACDC) at NATO Headquarters, strengthens dialogue among Allies, assesses risks to Allied populations, forces and territories, and supports chemical, biological, radiological or nuclear defence efforts.

- NATO is strengthening its capabilities to defend against chemical, biological, radiological or nuclear (CBRN) attacks, including terrorism and warfare.

- NATO conducts training and exercises designed to test interoperability and prepare forces to operate in a CBRN environment.

# OPENING UP OF ALLIANCE MEMBERSHIP

## The enlargement process

As we know provision for the enlargement of NATO is made in Article 10 of the North Atlantic Treaty. This is the basis of the open door policy adopted by NATO regarding the accession of new member countries. Decisions on the extension of invitations to potential new member countries to begin accession talks are taken jointly by all the existing members.

NATO's door remains open to any European country in a position to undertake the commitments and obligations of membership, and contribute to security in the Euro-Atlantic area. Since 1949, NATO's membership has increased from 12 to 30 countries through eight rounds of enlargement. Currently, three partner countries have declared their aspirations to NATO membership: Bosnia and Herzegovina, Georgia and Ukraine. The Republic of North Macedonia became the latest country to join the Alliance on 27 March 2020.

Remember that the roots of the changes which transformed the political map of Europe at the end of the 1980s can be traced to a number of developments during the 1960s and 1970s. Three events stand out in particular: the adoption by the Alliance, in December 1967, of the Harmel doctrine based on the parallel policies of maintaining adequate defence while seeking a relaxation of tensions in East-West relations and working towards solutions to the underlying political problems dividing Europe; the introduction by the Government of the Federal Republic of Germany in 1969 of Chancellor Willy Brandt's Ostpolitik, designed to bring about a more positive relationship with Eastern European countries and the Soviet Union within the constraints imposed by their governments' domestic policies and actions abroad; and the adoption of the Helsinki Final Act in August 1975, which established new standards and codes of conduct with regard to human rights issues and introduced measures to increase mutual confidence between East and West.

A series of similarly important events marked the course of East-West relations in the 1980s. These included NATO's deployment of intermediate-range nuclear forces (INF) in Europe following the December 1979 double-track decision on nuclear modernisation and arms control; the Washington Treaty, signed in December 1987, which brought about the elimination of US and Soviet land-based INF missiles on a global basis; early signs of change in Eastern Europe associated with the emergence and recognition, despite later setbacks, of the "Solidarity" independent trade union movement in Poland in August 1980; the consequences of the December 1979 Soviet invasion of Afghanistan and the ultimate withdrawal of Soviet forces from Afghanistan in February 1989; and, finally, the March 1985 nomination of Mikhail Gorbachev as General Secretary of the Soviet Communist Party, and his bold moves towards perestroika (restructuring) and glasnost (openness).

In the framework of the Conference on Security and Cooperation in Europe (CSCE) in March 1989, promising new arms control negotiations opened in Vienna between the 23 countries of NATO and the Warsaw Treaty Organization on reductions in conventional forces in Europe (CFE). The NATO Summit Meeting held in Brussels against this backdrop at the end of May 1989 was of particular significance. Members recognised the changes that were underway in the Soviet Union as well as in other Eastern European countries and outlined the Alliance's approach to overcoming the division of Europe and achieving its long-standing objective of shaping a just and peaceful European order. They reiterated the continuing need for credible and effective deterrent forces and an adequate defence, and set forth a broad agenda for expanded East-West cooperation.

So, developments of major significance for the entire European continent and for international relations as a whole continued as the year progressed. By the end of 1989 and during the early weeks of 1990, considerable progress was made towards the reform of the political and economic systems of Poland and Hungary. In the German Democratic Republic, Bulgaria, Czechoslovakia and Romania, steps were taken towards freedom and democracy which went far beyond expectations.

The promise held out for over 40 years to bring an end to the division of Europe, and with it an end to the division of Germany, took on real meaning with the fall of the Berlin Wall in November 1989. Beyond its fundamental symbolism, it opened the path to rapid and dramatic progress in most Central and Eastern European countries.

Thus, within less than a year, on 3 October 1990, the unification of the two German states took place with the backing of the international community and the acquiescence of the Soviet government, on the basis of an international treaty and the democratic will of the German people as a whole. Within just a few years, a number of Central and Eastern European countries had established membership of NATO as their principal foreign policy goal despite the negative image of the Alliance portrayed by the Soviet Union and Warsaw Pact governments during the Cold War.

NATO extended the hand of friendship to its former adversaries and initiated a process of dialogue and cooperation at its July 1990 London Summit. In December 1991, it created a joint forum for multilateral consultation and cooperation in the form of the North Atlantic Cooperation Council (NACC), and in January 1994 the Partnership for Peace (PfP) programme was launched to provide a framework for bilateral cooperation with each country on an individual basis. The NACC was replaced by the Euro-Atlantic Partnership Council (EAPC) in May 1997, which has since provided the overall political framework for cooperation between NATO and its Partner countries.

Within a short space of time, all the countries involved had responded positively to these successive initiatives and had begun participating in practical cooperation programmes. Several countries also began to seek support for their future accession to the North Atlantic Treaty.

In 1994, the Alliance recognised the need for a considered response, framed in terms of its overall objectives and long-term intentions for extending cooperation further afield and laying the basis for peace and stability throughout the Euro-Atlantic area.

NATO leaders stated that they "expect and would welcome NATO expansion that would reach to democratic states to our East" at the January 1994 Brussels Summit. They reaffirmed that the Alliance was open to membership for other European states in a position to further the principles of the North Atlantic Treaty and contribute to security in the North Atlantic area.

This way, practical steps were taken to move the process forward in a manner that ensured Alliance goals and policies would not be compromised and also reassured Russia and other countries that the process would pose no threat to them. The Alliance needed to demonstrate that, on the contrary, extending the sphere of stability in the Euro-Atlantic area would enhance their own security and would be in their interests.

Study on NATO Enlargement:

Hence, in 1995, the Alliance undertook a Study on NATO Enlargement to examine the "why and how" of future admissions into the Alliance. The results of the Study were shared with interested Partner countries and made public. With regard to the "why" of NATO enlargement, the study concluded that, with the end of the Cold War and the disappearance of the Warsaw Treaty Organization, there was both a need for and a unique opportunity to build improved security in the whole Euro-Atlantic area, without recreating new dividing lines.

In this sense, the study further concluded that enlargement of the Alliance would contribute to enhanced stability and security for all countries in the Euro-Atlantic area by encouraging and supporting democratic reforms, including the establishment of civilian and democratic control over military forces, fostering patterns and habits of cooperation, consultation and consensus-building characteristic of relations among members of the Alliance, and promoting good-neighbourly relations. It would increase transparency in defence planning and military budgeting, thereby reinforcing confidence among states, and would reinforce the overall tendency toward closer integration and cooperation in Europe. The study also concluded that enlargement would strengthen the Alliance's ability to contribute to European and international security.

With regard to the "how" of enlargement, the study confirmed that any future extension of Alliance membership would be through accession of new member states to the North Atlantic Treaty in accordance with its Article 10. Once admitted, new members would enjoy all the rights and assume all the obligations of membership. They would need to accept and conform to the principles, policies and procedures adopted by all the members of the Alliance at the time they joined. The willingness and ability to meet such commitments would be a critical factor in any decision taken by the Alliance to invite a country to join.

And other conditions were stipulated, including the need for candidate countries to settle ethnic disputes or external territorial disputes by peaceful means before they could become members and to treat minority populations in accordance with guidelines established by the Organization for Security and Co-operation in Europe.

The ability of candidate countries to contribute militarily to collective defence and to peacekeeping operations would also be a factor. Ultimately, the study concluded that Allies would decide by consensus whether to invite additional countries to join, basing their decision on their judgement at the time as to whether the membership of a specific country would contribute to security and stability in the Euro-Atlantic area.

NATO accession:

In order to explore the issues that had been raised in the 1995 Study on NATO Enlargement, NATO decided to conduct "intensified dialogues" with each of the countries that had declared their interest in joining the Alliance. Intensified dialogues were first launched with the Czech Republic, Hungary and Poland in early 1997, in the run-up to NATO's first post-Cold War round of enlargement in 1999. As early as July 1997, Allied heads of state and government were able to invite these three countries to begin accession talks. Accession protocols were signed in December 1997 and were then ratified by all 16 NATO countries. The three countries acceded to the Treaty, thereby becoming members of NATO in March 1999.

Allied heads of state and government invited Bulgaria, Estonia, Latvia, Lithuania, Romania, Slovakia and Slovenia to begin accession talks at the Prague Summit in November 2002. All seven of these countries had previously been participants in the Membership Action Plan (MAP). The procedures followed both by the existing NATO members and by the invited countries during the next twelve months illustrate the accession process that would apply to future member countries. Protocols of accession were signed by the foreign ministers of the invited countries at NATO Headquarters on 26 March 2003. By the end of April 2004, all Alliance member countries had notified the government of the United States of their acceptance of the protocols, in accordance with the North Atlantic Treaty, and on 2 March 2004, the NATO Secretary General formally invited the seven countries to become members. At a ceremony in Washington, DC, on 29 March 2004, each country deposited its formal instruments of accession, as prescribed by the Treaty, thereby legally and formally becoming a member country of the Alliance.

Between the moment when the seven were invited to start accession talks and the projected signing of accession protocols, ratification and membership, the invited countries were involved to the maximum extent in Alliance activities and continued to benefit from participation in the Membership Action Plan. Each of the invited countries also presented a timetable for necessary reforms to be carried out before and after accession in order to enhance their contribution to the Alliance.

A newly constructed extension to NATO Headquarters in Brussels was inaugurated by the Secretary General on 17 March 2004, providing accommodation for the delegations of the new member countries. On 2 April 2004, following the ceremonial raising of the flags of the new members outside NATO Headquarters, the first formal meeting of the North Atlantic Council with the participation of 26 member countries was held.

And by 1999, NATO Heads of State and Government came to realise that, with NATO's enlargement and transformation, the facilities no longer met the requirements of the Alliance. They agreed to construct a new Headquarters situated across the road from the existing Headquarters, Boulevard Léopold III, Brussels. The construction of the building was finalised in 2017 and the move took place in 2018.

Ultimately, once the Allies have decided to invite a country to become a member of NATO, they officially invite the country to begin accession talks with the Alliance. This is the first step in the accession process on the way to formal membership. The major steps in the process are:

1. Accession talks with a NATO team

These talks take place at NATO Headquarters in Brussels and bring together teams of NATO experts and representatives of the individual invitees.

Their aim is to obtain formal confirmation from the invitees of their willingness and ability to meet the political, legal and military obligations and commitments of NATO membership, as laid out in the Washington Treaty and in the Study on NATO Enlargement.

The talks take place in two sessions with each invitee. In the first session, political and defence or military issues are discussed, essentially providing the opportunity to establish that the preconditions for membership have been met.

The second session is more technical and includes discussion of resources, security, and legal issues as well as the contribution of each new member country to NATO's common budget. This is determined on a proportional basis, according to the size of their economies in relation to those of other Alliance member countries.

Invitees are also required to implement measures to ensure the protection of NATO classified information, and prepare their security and intelligence services to work with the NATO Office of Security.

The end product of these discussions is a timetable to be submitted by each invitee for the completion of necessary reforms, which may continue even after these countries have become NATO members.

2. Invitees send letters of intent to NATO, along with timetables for completion of reforms

In the second step of the accession process, each invitee country provides confirmation of its acceptance of the obligations and commitments of membership in the form of a letter of intent from each foreign minister addressed to the NATO Secretary General. Together with this letter they also formally submit their individual reform timetables.

3. Accession protocols are signed by NATO countries

NATO then prepares Accession Protocols to the Washington Treaty for each invitee. These protocols are in effect amendments or additions to the Treaty, which once signed and ratified by Allies, become an integral part of the Treaty itself and permit the invited countries to become parties to the Treaty.

4. Accession protocols are ratified by NATO countries

The governments of NATO member states ratify the protocols, according to their national requirements and procedures. The ratification procedure varies from country to country. For example, the United States requires a two-thirds majority to pass the required legislation in the Senate. Elsewhere, for example in the United Kingdom, no formal parliamentary vote is required.

5. The Secretary General invites the potential new members to access to the North Atlantic Treaty

Once all NATO member countries notify the Government of the United States of America, the depository of the Washington Treaty, of their acceptance of the protocols to the North Atlantic Treaty on the accession of the potential new members, the Secretary General invites the new countries to access to the Treaty.

6. Invitees access to the North Atlantic Treaty in accordance with their national procedures

7. Upon depositing their instruments of accession with the US State Department, invitees formally become NATO members


**The membership action plan**

The Membership Action Plan (MAP) is a NATO program of advice, assistance and practical support tailored to the individual needs of countries wishing to join the Alliance. Participation in the MAP does not prejudge any decision by the Alliance on future membership. Bosnia and Herzegovina is currently participating.

Countries participating in the MAP submit individual annual national programs on their preparations for possible future membership. These cover political, economic, defence, resource, security and legal aspects.

The MAP process provides a focused and candid feedback mechanism on aspirant countries' progress on their programs. This includes both political and technical advice, as well as annual meetings between all NATO members and individual aspirants at the level of the North Atlantic Council to assess progress, on the basis of an annual progress report. A key element is the defence planning approach for aspirants, which includes elaboration and review of agreed planning targets.

Throughout the year, meetings and workshops with NATO civilian and military experts in various fields allow for discussion of the entire spectrum of issues relevant to membership.

The MAP was launched in April 1999 at the Alliance's Washington Summit to help countries aspiring to NATO membership in their preparations. The process drew heavily on the experience gained during the accession process of the Czech Republic, Hungary and Poland, which became members in the Alliance's first post-Cold War round of enlargement in 1999.

Participation in the MAP

Participation in the MAP helped prepare the seven countries that joined NATO in the second post-Cold War round of enlargement in 2004 (Bulgaria, Estonia, Latvia, Lithuania, Romania, Slovakia and Slovenia) as well as Albania and Croatia, which joined in April 2009. Montenegro, which joined the MAP in December 2009, became a member of the Alliance in June 2017. The Republic of North Macedonia, which had been participating in the MAP since 1999, joined NATO in March 2020.

Currently, Bosnia and Herzegovina is participating in the MAP, having been invited to do so in 2010. At the time, Allied foreign ministers called on the authorities in Bosnia and Herzegovina to resolve a key issue concerning the registration of immovable defence property to the state . At their meeting in December 2018, foreign ministers decided that NATO is ready to accept the submission of Bosnia and Herzegovina's first Annual National Program under the MAP. The registration of immovable defence property to the state remains essential.

## PARTNERSHIP AND COOPERATION

The Allies seek to contribute to the efforts of the international community in projecting stability and strengthening security outside NATO territory. One of the means to do so is through cooperation and partnerships. Over more than 25 years, the Alliance has developed a network of partnerships with non-member countries from the Euro-Atlantic area, the Mediterranean and the Gulf region, and other partners across the globe. NATO pursues dialogue and practical cooperation with these nations on a wide range of political and security-related issues. NATO's partnerships are beneficial to all involved and contribute to improved security for the broader international community.

In a schematic way we can point out that:

1) Partners are part of many of NATO's core activities, from shaping policy to building defence capacity, developing interoperability and managing crises.

2) NATO's programs also help partner nations to develop their own defence and security institutions and forces.

3) In partnering with NATO, partners can:

- share insights on areas of common interest or concern through political consultations and intelligence-sharing;

- gain access to advice and support as they reform and strengthen defence institutions and capacities;

- participate in a rich menu of education, training and consultation events (over 1,200 events a year are open to partners through a Partnership Cooperation Menu);

- prepare together for future operations and missions by participating in exercises and training;
- contribute to current NATO-led operations and missions;

- share lessons learned from past operations and develop policy for the future;

- work together with Allies on research and capability development.

4) Through partnership, NATO and partners also pursue a broad vision of security:

- integrating gender perspectives into security and defence;

- fighting against corruption in the defence sector;

- enhancing efforts to control or destroy arms, ammunition and unexploded ordnance;

- advancing joint scientific projects.

5) Partnership has evolved over the years, to encompass more nations, more flexible instruments, and new forms of cooperation and consultation.

Dialogue and cooperation with partners can make a concrete contribution to enhance international security, to defend the values on which the Alliance is based, to NATO's operations, and to prepare interested nations for membership.

In both regional frameworks and on a bilateral level, NATO develops relations based on common values, reciprocity, mutual benefit and mutual respect.

In the Euro-Atlantic area, the 30 Allies engage in relations with 20 partner countries through the Euro-Atlantic Partnership Council and the Partnership for Peace - a major program of bilateral cooperation with individual Euro-Atlantic partners. Among these partners, NATO has developed specific structures for its relationships with Russia, Ukraine and Georgia.

NATO is developing relations with the seven countries on the southern Mediterranean rim through the Mediterranean Dialogue, as well as with four countries from the Gulf region through the Istanbul Cooperation Initiative.

NATO also cooperates with a range of countries which are not part of these regional partnership frameworks. Referred to as "partners across the globe", they include Afghanistan, Australia, Colombia, Iraq, Japan, the Republic of Korea, Mongolia, New Zealand and Pakistan.

NATO has also developed flexible means of cooperation with partners, across different regions. NATO can work with so-called "30+n" groups of partners, where partners are chosen based on a common interest or theme. At the 2014 Wales Summit, NATO introduced the possibility of "enhanced opportunities" for certain partners to build a deeper, more tailor-made bilateral relationship with NATO.

At the same time, Allied leaders launched the "Interoperability Platform", a permanent format for cooperation with partners on the interoperability needed for future crisis management and operations.

Under NATO's partnership policies, the strategic objectives of NATO's partner relations are to:

- Enhance Euro-Atlantic and international security, peace and stability;

- Promote regional security and cooperation;

- Facilitate mutually beneficial cooperation on issues of common interest, including international efforts to meet emerging security challenges;

- Prepare interested eligible nations for NATO membership;

- Promote democratic values and institutional reforms, especially in the defence and security sector;

- Enhance support for NATO-led operations and missions;

- Enhance awareness of security developments including through early warning, with a view to preventing crises;

- Build confidence and achieve better mutual understanding, including about NATO's role and activities, in particular through enhanced public diplomacy.

That said, each partner determines, with NATO, the pace, scope, intensity and focus of their partnership with NATO, as well as individual objectives. This is often captured in a document setting goals for the relationship, which is to be regularly reviewed. However, many of NATO's partnership activities involve more than one partner at a time.

**Euro-Atlantic Partnership**

The Alliance seeks to foster security, stability and democratic transformation across the Euro-Atlantic area by engaging in partnership through dialogue and cooperation with non-member countries in Europe, the Caucasus and Central Asia. The Euro-Atlantic Partnership is underpinned by two key mechanisms: the Euro-Atlantic Partnership Council (EAPC) and the Partnership for Peace (PfP) program.

As basic ideas we will say that:

a. The Euro-Atlantic Partnership brings together Allies and partner countries from Europe, the Caucasus and Central Asia for dialogue and consultation.

b. The 50-nation Euro-Atlantic Partnership Council (EAPC) is a multilateral forum for dialogue and consultation, and provides the overall political framework for cooperation between the 30 NATO members and 20 partner countries.

c.  The Partnership for Peace (PfP), launched in 1994, facilitates practical bilateral cooperation between individual partner countries and NATO, tailored according to the specific ambitions, needs and abilities of each partner

d.  NATO's 2010 Strategic Concept identifies the EAPC and PfP as central to the Allies 'vision of a Europe whole, free and at peace.

e.  As early as 1991, NATO had set up a forum to institutionalise relations with countries of the former Soviet Union and Warsaw Pact, called the North Atlantic Cooperation Council (replaced by the EAPC in 1997).

NATO's new Strategic Concept, which was approved at the Lisbon Summit in November 2010, states that the EAPC and the PfP program are central to the Allies' vision of a Europe whole, free and at peace. Three priorities underpin cooperation with partners:

-   Dialogue and consultations;

-   Building capabilities and strengthening interoperability; and

-   Supporting reform.

Activities under the EAPC and PfP are set out in the Euro-Atlantic Partnership Work Plan. This is a catalog of around

1600 activities covering over 30 areas of cooperation, ranging from arms control, through language training, foreign and security policy, and military geography.

The EAPC and the PfP program have steadily developed their own dynamic, as successive steps have been taken by NATO and its partner countries to extend security cooperation, building on the partnership arrangements they have created.

As NATO has transformed over the years to meet the new challenges of the evolving security environment, partnership has developed along with it. Today, partner countries are engaged with NATO in tackling 21st century security challenges, including terrorism and the proliferation of weapons of mass destruction.

The ways and means of cooperation developed under NATO's Euro-Atlantic Partnership have proven to be of mutual benefit to Allies and partners, and have helped promote stability. The mechanisms and programs for cooperation developed under EAPC/PfP are also being used as the basis to extend cooperation to other non-member countries beyond the Euro-Atlantic area.

Partners are expected to fund their own participation in cooperation programs. However, NATO supports the cost of individual participation of some nations in specific events, and may also support the hosting of events in some partner countries.

The Euro-Atlantic Partnership is about more than practical cooperation: it is also about values. Each partner country signs the PfP Framework Document. In doing so, partners commit to:

- respect international law, the United Nations Charter, the Universal Declaration of Human Rights, the Helsinki Final Act, and international disarmament and arms control agreements;

- refrain from the threat or use of force against other states;

- settle disputes peacefully.

The Framework Document also enshrines a commitment by the Allies to consult with any partner country that perceives a direct threat to its territorial integrity, political independence or security - a mechanism which, for example, Albania and the Republic of North Macedonia made use of during the Kosovo crisis. This commitment was also included as a provision in the 2009 Declaration to Complement the NATO-Ukraine Charter - in March 2014, with its independence and territorial integrity under threat, Ukraine invoked the provision and requested a meeting with Allies in the format of the NATO- Ukraine Commission.

See the years, 34 countries have joined the Euro-Atlantic Partnership. A number of these have since become NATO member states, through four rounds of NATO enlargement. This has changed the balance between Allies and partners in the EAPC/PfP: since March 2004, there have been more Allies than partners.

The remaining partners are a very diverse group, with different goals and ambitions with regard to their cooperation with NATO. They include Eastern and South-eastern European countries, the countries of the South Caucasus and Central Asia, and Western European states.

Some partners are in the process of reforming their defence structures and capabilities. Others are able to contribute significant forces to NATO-led operations and wish to further strengthen interoperability, and can also offer fellow partner countries advice, training and assistance in various areas. Other partners are interested in using their cooperation with NATO in order to prepare for membership in the Alliance.

Evolution of the Euro-Atlantic Partnership:

November 1989 saw the fall of the Berlin Wall, signaling the end of the Cold War. Within a short period, the remarkable pace of change in Central and Eastern Europe left NATO faced with a new and very different set of security challenges.

Allied leaders responded at their summit meeting in London, in July 1990, by extending a "hand of friendship" across the old East-West divide and proposing a new cooperative relationship with all the countries of Central and Eastern Europe.

This sea-change in attitudes was enshrined in a new strategic concept for the Alliance, issued in November 1991, which adopted a broader approach to security. Dialogue and cooperation would be essential parts of the approach required to manage the diversity of challenges facing the Alliance. The key goals were now to reduce the risk of conflict arising out of misunderstanding or design and to better manage crises affecting the security of the Allies; to increase mutual understanding and confidence among all European states; and to expand the opportunities for genuine partnership in dealing with common security problems.

The scene was set for the establishment in December 1991 of the North Atlantic Cooperation Council (NACC), a forum to bring together NATO and its new partner countries to discuss issues of common concern.

NACC consultations focused on residual Cold War security concerns such as the withdrawal of Russian troops from the Baltic States. Political cooperation was also launched on a number of security and defence-related issues.

The NACC broke new ground in many ways. However, it focused on multilateral, political dialogue and lacked the possibility of each partner country developing individual cooperative relations with NATO.

Deepening partnership:

This changed in 1994 with the launch of the Partnership for Peace (PfP), a major program of practical bilateral cooperation between NATO and individual partner countries, which represented a significant leap forward in the cooperative process.

And, in 1997, the Euro-Atlantic Partnership Council (EAPC) was created to replace the NACC and to build on its achievements, paving the way for the development of an enhanced and more operational partnership.

The EAPC and the PfP program have steadily developed their own dynamic, as successive steps have been taken by NATO and its partner countries to extend security cooperation, building on the partnership arrangements they have created.

Further initiatives have been taken to deepen cooperation between Allies and PfP partners at successive summit meetings in Madrid (1997), Washington (1999), Prague (2002), Istanbul (2004), Riga (2006), Bucharest (2008) and Lisbon ( 2010).

The 2010 Strategic Concept, adopted at Lisbon, stresses that cooperative security constitutes one of the Alliance's core tasks, together with collective defence and crisis management. It states that "The Alliance will actively engage to enhance international security, through partnership with relevant countries and other international organisations (…)". It also refers specifically to the EAPC and PfP as "central to our vision of Europe whole, free and in peace".

In 2011, when NATO foreign ministers met in Berlin, they approved a more efficient and flexible partnership policy, designed to streamline NATO's partnership tools in order to open all cooperative activities and exercises to all partners and to harmonise NATO's partnership programs.

Because of this, PfP activities have been opened up to other partnership frameworks and, vice-versa, PfP partners have been able to participate in activities hosted by the other cooperative frameworks.

**Partnership for Peace program**

The Partnership for Peace (PfP) is a program of practical bilateral cooperation between individual Euro-Atlantic partner countries and NATO. It allows partners to build up an individual relationship with NATO, choosing their own priorities for cooperation.

Five basic ideas in this cooperation are:

a) Based on a commitment to democratic principles, the purpose of the Partnership for Peace is to increase stability, diminish threats to peace and build strengthened security relationships between NATO and non-member countries in the Euro-Atlantic area.

b) The PfP was established in 1994 to enable participants to develop an individual relationship with NATO, choosing their own priorities for cooperation, and the level and pace of progress.

c) Activities on offer under the PfP program touch on virtually every field of NATO activity.

d) Since April 2011, all PfP activities and exercises are in principle open to all NATO partners, be they from the Euro-Atlantic region, the Mediterranean Dialogue, the Istanbul Cooperation Initiative or global partners.

e) Currently, there are 20 countries in the Partnership for Peace program.

Activities on offer under the PfP program touch on virtually every field of NATO activity, including defence-related work, defence reform, defence policy and planning, civil-military relations, education and training, military-to-military cooperation and exercises, civil emergency planning and disaster response, and cooperation on science and environmental issues.

Over the years, a range of PfP tools and mechanisms have been developed to support cooperation through a mix of policies, programs, action plans and arrangements. At the Lisbon Summit in November 2010, as part of a focused reform effort to develop a more efficient and flexible partnership policy, Allied leaders, decided to take steps to streamline NATO's partnership tools in order to open all cooperative activities and exercises to partners and to harmonise partnership programs.

The new partnerships policy approved by Allied foreign ministers in Berlin in April 2011 opened all cooperative activities and exercises as well as some programs that were previously offered only to PfP partners to all partners, whether they be Euro-Atlantic partners, countries participating in the Mediterranean Dialogue and the Istanbul Cooperation Initiative, or global partners.

The Euro-Atlantic Partnership Council provides the overall political framework for NATO's cooperation with Euro-Atlantic partners and the bilateral relationships developed between NATO and individual partner countries within the Partnership for Peace program.

Partner countries choose individual activities according to their ambitions and abilities. These are put forward to NATO in what is called a Presentation Document.

An Individual Partnership and Cooperation Program (previously called the Individual Partnership Program) is then jointly developed and agreed between NATO and each partner country. These two-year programs are drawn up from an extensive menu of activities, according to each country's specific interests and needs. All partners have access to the Partnership and Cooperation Menu, which comprises some 1,600 activities.

Some countries choose to deepen their cooperation with NATO by developing Individual Partnership Action Plans (IPAPs). Developed on a two-year basis, such plans are designed to bring together all the various cooperation mechanisms through which a partner country interacts with the Alliance, sharpening the focus of activities to better support their domestic reform efforts.


**Mediterranean Dialogue**

NATO's Mediterranean Dialogue (MD) was initiated in 1994 by the North Atlantic Council. It currently involves seven non-NATO countries of the Mediterranean region: Algeria, Egypt, Israel, Jordan, Mauritania, Morocco and Tunisia.

The Dialogue reflects the Alliance's view that security in Europe is closely linked to security and stability in the Mediterranean. It is an integral part of NATO's adaptation to the post-Cold War security environment, as well as an important component of the Alliance's policy of outreach and cooperation.

The Mediterranean Dialogue's overall aim is to:

contribute to regional security and stability
achieve better mutual understanding
dispel any misconceptions about NATO among Dialogue countries
The Mediterranean Dialogue is based upon the twin pillars of political dialogue and practical cooperation.

The Mediterranean Cooperation Group (MCG), established at the Madrid Summit in July 1997 under the supervision of the North Atlantic Council (NAC), had the overall responsibility for the Mediterranean Dialogue, until it was replaced in 2011 by the Political and Partnerships Committee, which is responsible for all partnerships. The Committee meets at the level of Political Counselors on a regular basis to discuss all matters related to the Dialogue including its further development.

Political consultations in the NATO+1 format are held on a regular basis both at Ambassadorial and working level. These discussions provide an opportunity for sharing views on a range of issues relevant to the security situation in the Mediterranean, as well as on the further development of the political and practical cooperation dimensions of the Dialogue.

Meetings in the NATO+7 format, including NAC + 7 meetings, are also held on a regular basis, in particular following the NATO Summit and Ministerial meetings, Chiefs-of-Defence meetings, and other major NATO events.

These meetings represent an opportunity for two-way political consultations between NATO and MD partners.

At the June 2004 Istanbul Summit, NATO's Heads of State and Government elevated the MD to a genuine partnership through the establishment of a more ambitious and expanded framework, which considerably enhanced both the MD's political and practical cooperation dimensions.

Since then, the constant increase in the number and quality of the NATO-MD political dialogue has recently reached a sustainable level. Consultations of the 29 Allies and seven MD countries take place on a regular basis on a bilateral and multilateral level, at Ministerial, Ambassadorial and working level formats. That has also included three meetings of the NATO and MD Foreign Ministers in December 2004, 2007 and 2008 in Brussels. Two meetings of NATO and MD Defence Ministers in 2006 and 2007 in Taormina, Italy and Seville, Spain. Ten meetings of the Chief of Defence of NATO and MD countries have also take place so far. The first ever NAC+7 meeting took place in Rabat, Morocco, in 2006 and, more recently, the first MD Policy Advisory Goup meeting with all seven MD partners took place in San Remo, Italy, on 15-16 September 2011.

The political dimension also includes visits by NATO Senior Officials, including the Secretary General and the Deputy Secretary General, to Mediterranean Dialogue countries. The main purpose of these visits is to conduct high-level political consultations with the relevant host authorities on the way forward in NATO's political and practical cooperation under the Mediterranean Dialogue.

The new Strategic Concept, which was adopted at the Lisbon Summit in November 2011, identifies cooperative security as one of three key priorities for the Alliance, and constitutes an opportunity to move partnerships to the next generation. Mediterranean Dialogue partners were actively involved in the debate leading to its adoption.

The Strategic Concept refers specifically to the MD, stating that: "We are firmly committed to the development of friendly and cooperative relations with all countries of the Mediterranean, and we intend to further develop the Mediterranean Dialogue in the coming years.

We will aim to deepen the cooperation with current members of the Mediterranean Dialogue and be open to the inclusion in the Mediterranean Dialogue of other countries of the region".

MD partners have reiterated their support for enhanced political consultations to better tailor the MD to their specific interests and to maintain the distinctive cooperation framework of the MD.

Measures of practical cooperation between NATO and Mediterranean Dialogue countries are laid down in an annual Work Program which aims at enhancing our partnership through cooperation in security-related issues.

The annual Work Program includes seminars, workshops and other practical activities in the fields of modernisation of the armed forces, civil emergency planning, crisis management, border security, small arms & light weapons, public diplomacy, scientific and environmental cooperation, as well as consultations on terrorism and the proliferation of weapons of mass destruction (WMD).

There is also a military dimension to the annual Work Program which includes invitations to Dialogue countries to observe (and in some cases participate) in NATO/PfP military exercises, attend courses and other academic activities at the NATO School (SHAPE) in Oberammergau (Germany ) and the NATO Defence College in Rome (Italy), and visit NATO military bodies.

The military program also includes port visits by NATO's Standing Naval Forces, on-site train-the-trainers sessions by Mobile Training Teams, and visits by NATO experts to assess the possibilities for further cooperation in the military field.

Furthermore, NATO+7 consultation meetings on the military program involving military representatives from NATO and the seven Mediterranean Dialogue countries are held twice a year.


**Istanbul Cooperation Initiative**

NATO's Istanbul Cooperation Initiative (ICI), launched at the Alliance's Summit in the Turkish city in June 2004, aims to contribute to long-term global and regional security by offering countries of the broader Middle East region practical bilateral security cooperation with NATO.´

ICI focuses on practical cooperation in areas where NATO can add value, notably in the security field. Six countries of the Gulf Cooperation Council were initially invited to participate. To date, four of these, Bahrain, Qatar, Kuwait and the United Arab Emirates, have joined. Saudi Arabia and Oman have also shown an interest in the Initiative.

Based on the principle of inclusiveness, the Initiative is, however, open to all interested countries of the broader Middle East region who subscribes to its aims and content, including the fight against terrorism and the proliferation of weapons of mass destruction.

Each interested country will be considered by the North Atlantic Council on a case-by-case basis and on its own merit. Participation of countries in the region in the Initiative as well as the pace and extent of their cooperation with NATO will depend in large measure on their individual response and level of interest.

The Initiative offers a menu of bilateral activities that countries can choose from, which comprises a range of cooperation areas, including:

I.    tailored advice on defence transformation, defence budgeting, defence planning and civil-military relations;

II.    military-to-military cooperation to contribute to interoperability through participation in selected military exercises and related education and training activities that could improve the ability of participating countries' forces to operate with those of the Alliance; and through participation in selected NATO and PfP exercises and in NATO-led operation on a case-by-case basis;

III.    cooperation in the fight against terrorism, including through intelligence-sharing;

IV.    cooperation in the Alliance's work on the proliferation of weapons of mass destruction and their means of delivery;

V.    cooperation regarding border security in connection with terrorism, small arms and light weapons and the fight against illegal trafficking;

VI.    civil emergency planning, including participating in training courses and exercises on disaster assistance.

Individual and Partnership Cooperation Program (IPCP) allow interested ICI countries and NATO to frame their practical cooperation in a more prospective and focused way, enabling interested countries to outline the main short and long-term objectives of their cooperation with the Alliance.

NATO recognises that dealing with today's complex new threats requires wide international cooperation and collective effort. That is why NATO has developed, and continues to develop, a network of partnerships in the security field.

The Initiative was preceded by a series of high level consultations conducted by the then Deputy Secretary General of NATO, Ambassador Minuto Rizzo, with six countries of the region in May, September and December 2004.

These were: Bahrain, Kuwait, Oman, Qatar, Saudi Arabia and the United Arab Emirates. During these consultations all of the countries expressed their interest in the Initiative.

ICI was launched at the Summit meeting of NATO Heads of State and Government in Istanbul, June 28, 2004. Following the Summit, from September to December 2004, the Deputy Secretary General of NATO paid a second round of visits to the six members of the Gulf Cooperation Council, to discuss the way ahead.

In the first three months of 2005, three countries: Bahrain, Kuwait and Qatar formally joined the ICI. In June 2005, the United Arab Emirates joined the Initiative.

The ICI has since developed both in the political and in the practical dimensions. While the political dialogue has evolved to include high-level meetings, the practical dimension was progressively enhanced through the opening of new partnership tools and activities as well as through the contribution of these countries to NATO-led operations. The multilateral dimension of the partnership also developed, with the first NAC + 4 meeting held in November 2008, followed by two other such meetings in 2009 and 2010.

Since the Istanbul Summit in 2004, an annual Menu of Practical Activities focusing on agreed priority areas has been opened to ICI countries and has been gradually enhanced. Whereas in 2007, the offer of cooperation to ICI countries included 328 activities / events, the 2011 Menu of Practical Activities now contains about 500 activities.

The NATO Training Cooperation Initiative (NTCI), launched at the 2007 Riga Summit, aims at complementing existing cooperation activities developed in the ICI framework through the establishment of a "NATO Regional Cooperation Course" at the NATO Defence College (NDC) in Rome, which consists in a ten-week strategic level course also focusing on current security challenges in the Middle East. ICI partners, as well as Saudi Arabia, actively participate in these courses.

The importance of public diplomacy has been underlined by ICI nations. High visibility events gave way to informal discussions on security related issues of common interest. The ICI Ambassadorial Conferences in Kuwait (2006), Bahrain (2008) and the United Arab Emirates (2009), which were attended by the Secretary General, the Deputy Secretary General and the 28 NATO Permanent Representatives, as well as by high-ranking officials, policymakers and opinion leaders from ICI countries, focused on discussing and addressing the perception of NATO in the Gulf, as well as ways to develop NATO-ICI partnership in its two dimensions. The fourth ICI Ambassadorial Conference took place in Qatar in February 2011 and focused on deepening NATO-ICI partnership.

The new Strategic Concept, adopted at the Lisbon Summit in November 2010, identifies cooperative security as one of three core tasks for the Alliance. It refers specifically to the ICI, and states: "We attach great importance to peace and stability in the Gulf region, and we intend to strengthen our cooperation in the Istanbul Cooperation Initiative. We will aim to develop a deeper security partnership with our Gulf partners and remain ready to welcome new partners in the Istanbul Cooperation Initiative".

With the approval of the new partnership policy at the meeting of NATO foreign ministers in Berlin in April 2011, all NATO partners will have access in principle to the same range and number of activities. This will dramatically expand the number of activities accessible to ICI countries.

ICI partners have also increasingly demonstrated their readiness to participate in NATO-led operations, acting as security providers. Today, several ICI partners actively contribute to the NATO ISAF operation in Afghanistan. Following the launch of Operation Unified Protector (OUP) in Libya, Qatar and the United Arab Emirates promptly provided air assets to the operation and were recognised as contributing nations, playing a key role in the success of the operation.

**Relations with partners across the globe**

NATO cooperates on an individual basis with a number of countries which are not part of its regional partnership frameworks. Referred to as "partners across the globe" or simply "global partners", they include Afghanistan, Australia, Colombia, Iraq, Japan, the Republic of Korea, Mongolia, New Zealand and Pakistan.

The five basic ideas in this section are:

a) The importance of reaching out to countries and organisations across the globe was underlined in the Strategic Concept adopted at the November 2010 Lisbon Summit.

b) Following the Lisbon Summit, NATO revised its partnership policy in April 2011 to better engage with partners.

c) Global partners now have access to the full range of activities NATO offers to all partners; each has developed an Individual Partnership Cooperation Program, choosing the areas where they wish to engage with NATO in a spirit of mutual benefit and reciprocity.

d) Most global partners contribute actively to NATO-led operations and missions.

e) NATO also consults with other non-member countries which have no bilateral program of cooperation (for example, China, India, Singapore, Indonesia, Malaysia) on issues such as counter piracy and countering narcotics in Afghanistan.

The support provided by global partners and other countries to NATO-led operations makes a significant contribution to international peace and security.

In the Balkans, Argentinean and Chilean forces have worked alongside NATO Allies to ensure security in Bosnia and Herzegovina. In Kosovo, Argentina has helped NATO personnel provide medical and social assistance to the local population and cooperated on peace agreement implementation since 1999.

In Afghanistan, a number of global partners such as Australia, the Republic of Korea and New Zealand, made important contributions to the NATO-led International

Security Assistance Force (ISAF) from 2003 to 2014. Many continue to work alongside Allies in the follow- on mission to train, advise and assist the Afghan security forces (Resolute Support). Other countries, such as Japan, have supported stabilisation efforts in Afghanistan without being involved in combat, by funding a large number of development projects and dispatching liaison officers.

Pakistan's support for the efforts of NATO and the international community in Afghanistan remains crucial to the success of the Alliance's mission, despite past differences. NATO remains committed to engaging with Pakistan in an effort to enlist support to stabilise Afghanistan.

The participation of partners in NATO-led peace-support operations is guided by the Political-Military Framework (PMF), which was developed for NATO-led operations. This framework provides for the involvement of contributing states in the planning and force generation processes through the International Coordination Center at Supreme Headquarters Allied Powers Europe (SHAPE). Building on lessons learned and reinforcing the habit of cooperation established through the Kosovo Force (KFOR) and ISAF, NATO Allies decided at the 2010 Lisbon Summit to review the PMF in order to update how NATO shapes decisions and works with partner countries on the operations and missions to which they contribute.

Typically, partner military forces are incorporated into operations on the same basis as are forces from NATO member countries. This implies that they are involved in the decision-making process through their association to the work of NATO committees, and through the posting of liaison officers in the operational headquarters or to SHAPE. They operate under the direct command of the operational commander through multinational divisional headquarters. Regular meetings of the North Atlantic Council, the Alliance's principal political decision-making body, with ambassadors, ministers and heads of state and government are held to discuss and review the operations.

NATO has maintained a dialogue with countries that are not part of its partnership frameworks, on an ad-hoc basis, since the 1990s. However, NATO's involvement in areas outside of its immediate region - including Afghanistan and Libya - has increased the need and opportunities for enhanced global interaction. Clearly, the emergence of global threats requires the cooperation of a wider range of countries to successfully tackle challenges such as terrorism, proliferation, piracy or cyber attacks. Dialogue with these countries can also help NATO avert crises and, when needed, manage an operation throughout all phases.

Since 1998, NATO has invited countries across the globe to participate in its activities, workshops, exercises and conferences. This decision marked a policy shift for the Alliance, allowing these countries to have access, through the case-by-case approval of the North Atlantic Council, to activities offered under NATO's structured partnerships. These countries were known as "Contact Countries".

Significant steps were taken at the 2006 Riga Summit to increase the operational relevance of NATO's cooperation with countries that are part of its structured partnership frameworks as well as other countries around the world. These steps, reinforced by decisions at the 2008 Bucharest Summit, defined a set of objectives for these relationships and created avenues for enhanced political dialogue, including meetings of the North Atlantic Council with ministers of the countries concerned, high-level talks, and meetings with ambassadors. In addition, annual work programs (then referred to as Individual Tailored Cooperation Packages of Activities) were further developed.

At the 2010 Lisbon Summit, Allies agreed to develop a more efficient and flexible partnership policy, in time for the meeting of Allied foreign ministers in Berlin in April 2011. To this end, they decided to:

I.    streamline NATO's partnership tools in order to open all cooperative activities and exercises to partners and to harmonise partnership programs;

II.   better engage with partners across the globe who contribute significantly to security and reach out to relevant partners to build trust, increase transparency and develop practical cooperation;

III.  develop flexible formats to discuss security challenges with partners and enhance existing fora for political dialogue; and

IV.   build on improvements in NATO's training mechanisms and consider methods to enhance individual partners' ability to build capacity.


**Partnership Interoperability Initiative**

The Partnership Interoperability Initiative (PII) was launched at the Wales Summit in 2014 to ensure that the deep connections built up between NATO and partner forces over years of operations will be maintained and deepened. In this way, partners can contribute to future crisis management, including NATO-led operations and, where applicable, to the NATO Response Force.

We can say that the basic ideas to take into account in this topic are:

a) NATO partners contribute to NATO-led operations and missions, as well as exercises, often significantly.

b) Partner forces need to be interoperable - able to operate together with NATO forces according to NATO standards, rules, procedures and using similar equipment.

c) At the 2014 Wales Summit, NATO launched the Partnership Interoperability Initiative (PII) to maintain and deepen the interoperability that has been developed with partners during NATO-led operations and missions over the last decades.

d) The PII underlined the importance of interoperability for all its partnerships and proposed new means to deepen cooperation with those partners that wished to be more interoperable with NATO.

e) As a result of the PII, NATO granted tailor-made "enhanced opportunities" for deeper cooperation with five partners: Australia, Finland, Georgia, Jordan and Sweden.

f) The PII also launched the "Interoperability Platform" (IP) to provide a wider group of partners with deeper access to cooperation on interoperability issues - currently 23 selected partners, who are interested and committed to deepening interoperability for future crises, participate in meetings of a number of NATO committees and bodies held in the IP format.

Partners can contribute to NATO-led operations and missions - whether through supporting peace by training security forces in the Western Balkans and Afghanistan, or monitoring maritime activity in the Mediterranean Sea or off the Horn of Africa - as well as NATO exercises. To be able to contribute effectively, partners need to be interoperable with NATO.

Interoperability is the ability to operate together using harmonised standards, doctrines, procedures and equipment. It is essential to the work of an alliance of multiple countries with national defence forces, and is equally important for working together with partners that wish to contribute in supporting NATO in achieving its tactical, operational and strategic objectives. Much of day-to-day cooperation in NATO - including with partners - is focused on achieving this interoperability.

The Partnership Interoperability Initiative (PII):

In 2014, Allied leaders responded to the need to maintain and enhance interoperability built up with partners during years of operations (including in Afghanistan and the Western Balkans), recognising the importance of maintaining interoperability with partners for future crisis management. NATO launched the Partnership Interoperability Initiative (PII), which aims to:

- re-emphasise the importance of developing interoperability with and for all partners, and of ensuring that all existing partnership interoperability programs are used to their full potential;

- enhance support for those partners that wish to maintain and enhance their interoperability, including through deeper cooperation and dialogue;

- offer enhanced opportunities for cooperation to those partners that provide sustained and significant force, capability or other contributions to the Alliance;

- underline that interoperability also needs to be a priority for NATO's relations with other international organisations with a role in international crisis management.

The PII recognised that deeper interoperability underpins and complements closer relations between NATO and partners. As partner nations 'contributions to NATO missions and operations as well as force pools became more ambitious and complex, they would benefit from a more tailor-made relationship to help sustain such contributions, based on specific" enhanced opportunities "for cooperation, including:

- regular, political consultations on security matters, including possibly at ministerial level;

- enhanced access to interoperability programs and exercises;

- sharing information, including on lessons learned;

- closer association of such partners in times of crisis and the preparation of operations.

Shortly after the 2014 Wales Summit, five partners were granted these "enhanced opportunities": Australia, Finland, Georgia, Jordan, and Sweden. Since then, each "Enhanced Opportunities Partner" (EOP) has taken forward this program of cooperation with NATO in a tailor-made manner, in areas of mutual interest for NATO and the partner concerned.

Interoperability for current and future military cooperation to tackle security challenges is a key focus of day-to-day work at NATO, including in a broad range of committees, working groups and expert communities. The PII recognised that if partners are to be interoperable to manage crises with NATO tomorrow, they need to work with NATO on interoperability issues today - and be part of those discussions.

This is why the PII launched a standing format for NATO-partner cooperation on interoperability and related issues: the Interoperability Platform (IP). The format cuts across traditional, geographical frameworks for cooperation, and brings together all partners that have contributed to NATO operations or have taken concrete steps to deepen their interoperability with NATO.

Participation in these programs and activities changes, so the North Atlantic Council - the Alliance's highest political decision-making body - adjusts participation every year. As of June 2017, 24 partners are members of the IP.

In this format, Allies and partners discuss projects and issues that affect interoperability for future crisis management, such as command and control systems, education and training, exercises or logistics.

Recognising the breadth and depth of work needed on interoperability, any NATO committee or body can meet in IP format, at different levels. It was launched by a meeting of defence ministers in IP format at the Wales Summit, and since then has met in a number of configurations at NATO Headquarters, including at the level of the North Atlantic Council, the Military Committee, the Partnerships and Cooperative Security Committee, the Operations Policy Committee, and technical groups such as the Conference of National Armaments Directors, the Command, Control and Consultation Board, the Civil Emergency Planning Committee and others.

At the Warsaw Summit in July 2016, the defence ministers of the IP nations will meet with their NATO counterparts to review progress since Wales.

The following 23 partners are part of the IP:

Armenia, Australia, Austria, Azerbaijan, Bahrain, Bosnia and Herzegovina, Finland, Georgia, Ireland, Japan, Jordan, Kazakhstan, Republic of Korea, Republic of Moldova, Mongolia, Morocco, New Zealand, Serbia, Sweden,

Switzerland, Tunisia, Ukraine , and the United Arab Emirates.

**Partnership tools**

NATO has developed a number of partnership tools and mechanisms to support cooperation with partner countries through a mix of policies, programs, action plans and other arrangements. Many tools are focused on the important priorities of interoperability and building capabilities, and supporting defence and security-related reform.

The three fundamental ideas in this area are:

a) A Partnership Cooperation Menu comprising approximately 1,400 activities is accessible to all NATO partners.

b) Several initiatives are open to all partners that allow them to cooperate with NATO mainly focusing on interoperability and building capacity, and supporting defence and security-related reform.

c) Partnership tools for deeper bilateral cooperation with individual partners in specific areas include, for instance, the Planning and Review Process, the Operational Capabilities Concept and the Individual Partnership Action Plans.

Each partner determines the pace, scope, intensity and focus of their partnership with NATO, as well as individual objectives. Bilateral (NATO-partner) cooperation documents set out the main objectives and goals of that partner's cooperation with NATO. There are three main types of bilateral partnership documents, set out below. Broadly speaking, the type of document chosen reflects the different nature and emphasis of the relationship:

I. The Individual Partnership and Cooperation Program (IPCP) is the standard document, developed usually every two years by the partner in close consultation with NATO staffs, and then approved by the North Atlantic Council (NAC) and the partner. It is open to all partners, and is modular in structure, adaptable to the interests and objectives of the partner and NATO.

II. The Individual Partnership Action Plan (IPAP), which partners can take up instead of IPCPs, offer partners the opportunity to deepen their cooperation with NATO and sharpen the focus on domestic reform efforts. Developed on a two-year basis, these plans include a wide range of jointly agreed objectives and targets for reforms on political issues as well as security and defence issues. IPAP prioritises and coordinates all aspects of the NATO-partner relationship, provides for an enhanced political dialogue and systematic support to democratic and defence and related security sector reform, including through an annual Allied assessment of progress in reforms undertaken by each participating partner.

III. The Annual National Program (ANP) is the most demanding document, focused on comprehensive democratic, security and defence reforms, developed annually by the partner in consultation with NATO.

The ANP is open to Membership Action Plan (MAP) nations, to track progress on the road to NATO membership; Georgia in the context of the NATO-Georgia Commission; and Ukraine in the context of the NATO-Ukraine Commission. Unlike the IPCP or IPAP, the ANP is a nationally owned document and is not agreed by the NAC. However, an annual assessment of progress in reforms is conducted by NATO staffs, agreed by the Allies, and discussed with each participating partner at NAC level.

Partner countries have made and continue to make significant contributions to the Alliance's operations and missions, whether it be supporting peace in the Western Balkans and Afghanistan, training national security forces in Iraq, monitoring maritime activity in the Mediterranean Sea, or helping protect civilians in Libya .

At the Wales Summit in September 2014, NATO leaders endorsed two important initiatives to reinforce the Alliance's commitment to the core task of cooperative security: the Partnership Interoperability Initiative and the Defence and Related Security Capacity Building Initiative. The first initiative was designed to reinforce NATO's ability to provide security with partners in future, through interoperability; while the second was more focused on helping partners provide for their own security, by strengthening their defences and related security capacity. A number of tools have been developed to assist partners in developing their own defence capacities and defence institutions, ensuring that partner forces are able to provide for their own security, capable of participating in NATO-led operations, and interoperable with Allies 'forces.

They include the following:

1. The Planning and Review Process (PARP) helps develop the interoperability and capabilities of forces which might be made available for NATO training, exercises and operations. Under PARP, Allies and partners, together negotiate and set planning targets with a partner country. Regular reviews measure progress.

2. In addition, PARP also provides a framework to assist partners to develop effective, affordable and sustainable armed forces as well as to promote wider defence and security sector transformation and reform efforts. It is the main instrument used to assess the implementation of defence-related objectives and targets defined under IPAPs. PARP is open to Euro-Atlantic partners on a voluntary basis and is open to other partner countries on a case-by-case basis, upon approval of the NAC.

3. The Operational Capabilities Concept (OCC) Evaluation and Feedback Program is used to develop and train partner land, maritime, air or Special Operations Forces that seek to meet NATO standards. This rigorous process can often take a few years, but it ensures that partner forces are ready to work with Allied forces once deployed. Some partners use the OCC as a strategic tool to transform their defence forces. The OCC has contributed significantly to the increasing number of partner

forces participating in NATO-led operations and the NATO Response Force.

4.  Exercising is key for maintaining, testing and evaluating readiness and interoperability, also for partners. NATO offers partners a chance to participate in the Military Training and Exercise Program (MTEP) to promote their interoperability. Through the MTEP, a five-year planning horizon provides a starting point for exercise planning and the allocation of resources.

5.  In addition, and on a case-by-case basis, Allies may invite partners to take part in crisis-management exercises that engage the NAC and ministries in participating capitals, and national political and military representation at NATO Headquarters, in consultations on the strategic management of crises during an exercise.

6.  Once a partner wishes to join a NATO-led operation, the Political-Military Framework (PMF) sets out principles and guidelines for the involvement of all partner countries in political consultations and decision-shaping, in operational planning and in command arrangements for operations to which they contribute.

7.  Several tools and programs have been developed to provide assistance to partner countries in their own efforts to transform defence and security-related structures and policies, and to manage the economic and social consequences of reforms. An important priority is to promote the development of effective

defence institutions that are under civil and democratic control.

8. In particular, since 2014, the Defence and Related Security Capacity Building (DCB) Initiative reinforces NATO's commitment to partners and helps project stability by providing support to nations requesting defence capacity assistance from NATO. It can include various types of support, ranging from strategic advice on defence and security sector reform and institution building, to development of local forces through education and training, or advice and assistance in specialised areas such as logistics or cyber defence.

9. The Building Integrity Initiative is aimed at promoting good practice, strengthening transparency, accountability and integrity to reduce the risk of corruption in the defence establishments of Allies and partners alike.

10. In addition, a Professional Development Program can be launched for the civilian personnel of defence and security establishments to strengthen the capacity for democratic management and oversight.

11. Through the Partnership Trust Fund policy, individual Allies and partners support practical demilitarisation projects and defence transformation projects in partner countries through individual Trust Funds.

NATO offers different means to access education, training and exercises, which can help partners to train and test personnel in the various areas relevant to their NATO partnerships:

I. Education and training in various areas is offered to decision-makers, military forces, civil servants and representatives of civil society through institutions such as the NATO School in Oberammergau, Germany; the NATO Defence College in Rome, Italy; and some 30 national Partnership Training and Education Centers.

II. NATO offers partners a Partnership Cooperation Menu (PCM) - an annual catalog which comprises, on average, some 1,400 education, training and other events for partners across 37 disciplines, held in more than 50 countries, which cater to the needs of around 10, 000 participants from partner countries. In addition to NATO bodies, Allies and partners can offer contributions to the PCM.

III. To support education and training for defence reform, the Defence Education Enhancement Programs (DEEPs) are tailored programs through which the Alliance advises partners on how to build, develop and reform educational institutions in the security, defence and military domain.

Wider cooperation:

A) The NATO Science for Peace and Security (SPS) Program promotes joint cooperative projects between Allies and partners in the field of security-related civil science and technology. Funding applications should address SPS key priorities -

these are linked to NATO's strategic objectives and focus on projects in direct support to NATO's operations, as well as projects that enhance defec capacity building and address other security threats.

B) Disaster response and preparedness is also an important area of cooperation with partners. The Euro-Atlantic Disaster Response Coordination Center (EADRCC) is a 24/7 focal point for coordinating disaster-relief and consequence management efforts among NATO and partner countries, and has guided consequence-management efforts in more than 45 emergencies, including fighting floods and forest fires, and dealing with the aftermath of earthquakes.

C) The principles of United Nations Security Council Resolution (UNSCR) 1325 and related Resolutions - that form the Women, Peace and Security agenda - were first developed into a NATO policy approved by Allies and partners in the Euro-Atlantic Partnership Council (EAPC) in 2007 .The Resolutions reaffirm the role of women in conflict and post-conflict situations and encourage greater participation of women and the incorporation of gender perspectives in peace and security efforts. Over the years, the policy has been updated, related action plans have been strengthened and more partner countries from across the globe have become associated with these efforts. Currently NATO's UNSCR 1325 coalition is the largest worldwide with 55 nations associated to the Action Plan. In practice, NATO has made significant progress in embedding gender perspectives within education, training and exercises, as well as the planning and execution of missions and operations, policies and guidelines.

## Defence and Related Security Capacity Building Initiative

The Defence and Related Security Capacity Building (DCB) Initiative reinforces NATO's commitment to partners and helps project stability by providing support to nations requesting assistance from NATO. DCB helps partners improve their defence and related security capacities, as well as their resilience, and, therefore, contributes to the security of the Alliance. It can include various types of support, ranging from strategic advice on defence and security sector reform and institution-building, to development of local forces through education and training, or advice and assistance in specialised areas such as logistics or cyber defence.

In this section the basic ideas are:

a) The DCB Initiative was launched in September 2014 at the NATO Summit in Wales.

b) The Initiative is demand-driven and tailored to the needs of the recipient nations by providing support which reinforces and exceeds what is offered through other existing programs.

c) The Initiative builds on NATO's extensive track record and expertise in advising, assisting, training and mentoring countries that require defence and related security capacity building support. It uses

NATO's unique defence expertise to provide and coordinate practical specialised support.

d) Good progress continues on the DCB packages for Georgia, Iraq, Jordan and the Republic of Moldova.

e) At the Brussels Summit in July 2018, Allies approved a DCB package for Tunisia.

f) The packages are implemented with the support of Allies and partners, who provide advisors, trainers and coordinators to work with the recipient countries, and help fund projects. A dedicated DCB Trust Fund is in place, since 2015, to provide financial support to the Initiative.

g) NATO has also received a request for DCB support from Libya.

NATO has been providing capacity-building through a number of partnership programs and also as part of its operations and missions. The DCB Initiative enhances this role by allowing NATO to undertake DCB activities in support of partner nations, other non-partner nations or other international organisations. Any NATO assistance is provided following a specific request by the recipient country (which is then thoroughly assessed and considered by the North Atlantic Council) and relies on mutual political commitment and local ownership.

If existing programs cannot accommodate the request, then the Alliance may consider offering a tailored set of assistance measures, a specific "DCB package". Five DCB packages have been launched thus far. Additionally, NATO has received a request for DCB support from Libya.

Georgia:

The DCB package for Georgia was agreed in 2014 at the Wales Summit and intensified in 2016 at the Warsaw Summit.

It is provided through the Substantial NATO-Georgia Package (SNGP), which includes support in a wide range of areas: the Joint Training and Evaluation Center, the Defence Institution Building School, a logistics capability, acquisitions, Special Operations Forces, intelligence-sharing and secure communications, military police, cyber defence, maritime security, aviation, air defence, strategic communications, crisis management and counter-mobility. The package also includes support and contributions to NATO exercises in Georgia that are open to partners.

Since 2014, many projects and advisory activities have been launched in support of the SNGP initiatives. One of the highlights was the inauguration of the NATO-Georgia Joint Training and Evaluation Center in August 2015 by Georgian leaders and the NATO Secretary General. The center is tasked with strengthening the capacities of the Georgian Armed Forces, as well as improving the interoperability of Georgian and Allied forces and contributing to regional security cooperation. It has conducted many activities since its establishment, and it will play an important role in the upcoming 2019 NATO-Georgia exercise. Another flagship initiative of the SNGP, the Defence Institution Building School, also continues to produce results, running specialised courses. Other SNGP initiatives also make progress. The strategic and operational planning initiative was completed in 2017.

At the Brussels Summit in July 2018, the Allies and Georgia declared they would further enhance cooperation, including through the next NATO-Georgia exercise in March 2019, which Allies will support with broad participation. NATO is moving ahead with the establishment of secure communications with Georgia and stepping up support in the area of counter-mobility. Dialogue is ongoing on hybrid threats and resilience, and cooperation in cyber defence may be enhanced to further strengthen interoperability.

The SNGP is currently supported by all Allies and two partners, who all together provide more than 40 experts, resident or frequently traveling to Georgia. A three-person Core Team in Tbilisi coordinates the implementation of the package.

Iraq:

The DCB package for Iraq was agreed in July 2015 following a request from the Iraqi Prime Minister. At the 2016 NATO Summit in Warsaw, NATO agreed to transfer the training and capacity-building activities inside Iraq based on the request of the Iraqi Prime Minister. The NATO Training and Capacity Building activity in Iraq currently conduct activities in the following areas: counter-improvised explosive devices (C-IED), explosive ordnance disposal and demining; civil-military planning support to operations; reform of the Iraqi security institutions; technical training on the maintenance of Soviet-era armored vehicles; military medicine and medical assistance; advice on security sector reform (SSR); and civil-military planning support to operations.

In-country training started in January 2017 with a "train-the-trainer" focus, aiming primarily to increase the training capacity of Iraq. The activities conducted range from multiple workshops on civilian-military cooperation, train-the-trainer courses to Iraqi instructors, to senior leader's seminars on C-IED. In the SSR area, NATO is providing advice to the Iraqi authorities on the transformation and good governance of the defence sector.

One of the key principles of NATO's capacity-building activities is to seek complementarity with other international actors. As such, NATO works closely with the Global Coalition to Defeat ISIL, the European Union, the United Nations and individual nations providing support to Iraq. One such example of this are the combined workshops conducted by the international community assisting Iraq's SSR, which was supported by NATO experts.

Responding to a request from the Government of Iraq for additional support in its efforts to stabilise the country and fight terrorism, at the Brussels Summit in July 2018, Allies decided to intensify the level of support for Iraq by launching a non-combat and capacity building mission in Iraq.

Building on current training activities, the NATO Training Mission Iraq will advise relevant Iraqi officials, primarily in the Ministry of Defence and the Office of the National Security Advisor, and train and advise instructors at professional military education institutions to help Iraq develop its capacity to build more effective national security structures and professional military education institutions.

Jordan:

The DCB assistance for Jordan builds upon the already extensive level of cooperation between NATO and Jordan through various partnership tools. The initial DCB package agreed in 2014 at the Wales Summit was revised and approved in 2017 reflecting the progress made and addressing the evolving security needs of the Jordanian Armed Forces.

The package focuses on the areas of information protection, cyber defence, military exercises, C-IED, strategic defence review, personnel management, logistics system, civil preparedness/crisis management and border security.

Activities are underway in elements of the package, ranging from courses for Jordanian personnel on C-IED to advice on strategy and capability development in other areas. The support provided on C-IED, cyber defence and exercises has been particularly fruitful. A Computer Emergency Response Team has been established for the Jordanian Armed Forces, which has a nation-wide responsibility. Jordan hosted successfully the NATO Regional Exercise 2017 (REGEX 2017), the first NATO exercise held in a Mediterranean Dialogue country.

The implementation of the package is supported, inter alia, by NATO's Science for Peace and Security Program in the areas of C-IED, cyber defence and border security, as well as the DCB Trust Fund projects, particularly in the areas of logistics (codification) and civil preparedness/crisis management.

Republic of Moldova:

Following the commitment made at the 2014 Wales Summit, the DCB package for the Republic of Moldova was launched in June 2015.

The package will be delivered in two phases. In phase one, which is currently underway, NATO is advising and assisting in the establishment of a national security strategy, national defence strategy, a military strategy and a force structure for Moldova. NATO brings defence reform experts to Moldova on a frequent basis to assist Moldovan authorities as they develop these key political and strategic-level directions and guidance for the defence sector and the development of the armed forces. In parallel to the defence sector reform, NATO has been providing support to Moldova in several specific areas, such as cyber defence, defence education, building integrity and the implementation of United Nations Security Council Resolution 1325 on Women, Peace and Security.

In phase two, NATO will continue to provide advice and will assist with specific elements of the transformation of Moldova's armed forces and relevant institutions.

Tunisia:

At the Brussels Summit in July 2018, in response to a request from the Tunisian authorities, Allies approved new DCB assistance measures designed to help further develop their defence capacities in the areas of cyber defence, countering improvised explosive devices, and promoting transparency in resource management . This DCB package will be implemented mainly through education and training activities and the exchange of expertise and best practices, in line with NATO standards.

Finally, the DCB Trust Fund was established in 2015 to provide financial support and resources to implement the DCB Initiatives. The Trust Fund allows Allies and partners to contribute, on a voluntary basis, to the implementation of projects developed in support of the packages. It has proven to be an important enabler to kick-start DCB activities. Since the establishment of the DCB Trust Fund, it has facilitated 15 projects and is currently supporting another seven projects.

**Contact Point Embassies in partner countries**

Since the early 1990s, NATO has developed a network of Contact Point Embassies (CPE) to support the Alliance's partnership and public diplomacy activities in countries participating in the Euro-Atlantic Partnership Council (EAPC), Partnership for Peace (PfP), Mediterranean Dialogue ( MD) and Istanbul Cooperation Initiative (ICI). Following the review of NATO's partnerships policy in April 2011, the network of CPEs has also been extended to other partners across the globe.

CPEs are a valuable tool which contribute to NATO's outreach efforts. In every partner country an embassy of one of the NATO member states serves as a contact point and operates as a channel for disseminating information about the role and policies of the Alliance. In addition to this public diplomacy role, the CPEs mandate has been extended to also include support - as required - for the implementation of other agreed activities with partners.

CPEs work closely with NATO's Public Diplomacy Division to provide information on the purpose and activities of the Alliance in the host country while also supporting the Political Affairs and Security Policy Division with its management of EAPC, PfP, MD and ICI policy.

CPEs are not NATO's diplomatic mission in the host country; However, they play an important role in disseminating information about the Alliance. CPEs identify key decision makers, opinion formers and public diplomacy opportunities within the country and coordinate with the Public Diplomacy Division on events. CPEs also inform individuals within the host country on how to apply for NATO fellowships and participate in scientific programs.

CPEs offer advice to NATO Headquarters on various project proposals as well as on an array of NATO-related issues within the host country, such as political discussions, debates and concerns and changes in public opinion. CPEs also assist with logistical support, political advice and briefings on relevant developments in the host country in preparation for visits to the country by the Secretary General, NATO International Staff and NATO forces. They also regularly liaise with other NATO member nation embassies in the host country to inform about NATO's agenda and involve them in NATO-related activities or events.

NATO's member countries volunteer the services of their embassies in partner countries to assume the duties of CPE for a period of two years. The final decision on the assignment of CPEs is taken by consensus in the North Atlantic Council - the principal political decision-making body within NATO. PDD coordinates the CPE network and liaises closely with each CPE.

# NATO'S RELATIONS WITH RUSSIA AND UKRAINE

Given the purpose for which NATO emerged and given the special role that Russia, and therefore Ukraine, has played and is playing, in the geopolitical and military context it is important to dedicate a special chapter to the Alliance's relationship with both countries.

## NATO and Russia

Let us start by saying that since the end of the Cold War, NATO member countries have regarded the development of a positive relationship and cooperation with Russia as a priority.

Over the years, much progress has been made in transforming old antagonisms based on ideological, political and military confrontation into an evolving and formally constituted partnership founded on common interests and continuing dialogue.

Although relations between the two sides have been strained by the attack on Ukraine by Russian-backed armed forces, NATO member states and Russia have been meeting regularly as equals in the NATO-Russia Council to consult on current security issues and develop practical cooperation in a wide range of areas of common interest.

While differences remain on some issues which may take some time to resolve, the driving force behind the new spirit of cooperation is the realisation that NATO member states and Russia share strategic priorities and face common challenges, such as the fight against terrorism and the proliferation of weapons of mass destruction.

The evolution of relations:

Remember that the ideological and political division of Europe ended in 1989 with the fall of the Berlin Wall. Following the disintegration of the Soviet Union and the Warsaw Pact in 1991, and with the upsurge of new security challenges in the post-Cold War environment, NATO began establishing new forms of dialogue and cooperation with the countries of central and eastern Europe and the member countries of the Commonwealth of Independent States (CIS).

Russia became a member of the North Atlantic Cooperation Council (which was replaced by the Euro-Atlantic Partnership Council in 1997) in 1991. And in 1994, it joined NATO's Partnership for Peace (PfP), a major programme of bilateral cooperation. In 1996, after the signature of the Dayton Peace Accord, Russia contributed troops and logistical support to the NATO-led peacekeeping force in Bosnia and Herzegovina.

It should also be remembered that cooperation in complex field conditions in the Balkans significantly reinforced mutual trust and strengthened the political will to take NATO-Russia cooperation to a new level. That transformation occurred in May 1997, with the signature of the Founding Act on Mutual Relations, Cooperation and Security, which provided the formal basis for NATO-Russia relations. It expressed the common goal of building a lasting peace and established the Permanent Joint Council (PJC) as a forum for consultation and cooperation. As a result,

NATO-Russia relations took on a concrete institutional dimension, in addition to the operational dimension.

Thus, in the years that followed, considerable progress was made in building mutual confidence and developing a programme of consultation and cooperation. However, lingering Cold War prejudices prevented the PJC from achieving its potential. In 1999, when differences arose over NATO's Kosovo air campaign, Russia suspended its participation in the PJC, which up to then had met on a regular basis at ambassadorial or ministerial level. Nevertheless, several activities continued without interruption, including peacekeeping in Bosnia and Herzegovina. Moreover, Russia played a key diplomatic role in resolving the Kosovo crisis and, in June 1999, when the NATO-led Kosovo Force was eventually deployed, Russian peacekeepers were a part of it.
From 1999 onwards, NATO-Russia relations started to improve significantly. When Lord Robertson became NATO Secretary General in October of that year, he committed himself to breaking the stalemate in NATO-Russia relations. Similarly, in 2000, upon his election as President of Russia, Vladimir Putin announced that he would work to rebuild relations with NATO in a spirit of pragmatism.

Several key events also accelerated this process. On 12 August 2000, the nuclear submarine Kursk sunk killing all 118 crewmen aboard, highlighting the urgent need for cooperation between NATO and Russia in responding to such tragic accidents. The terrorist attacks on the United States of 11 September 2001 also served as a stark reminder that concerted international action was needed to effectively tackle terrorism and other new security threats. In the immediate aftermath of the terrorist attacks, Russia opened its airspace for the international coalition's campaign in Afghanistan and shared intelligence to support the anti-terrorist coalition.

High-level contacts between NATO and Russia in the following months, including two meetings of Lord Robertson with President Putin and a meeting of Allied and Russian foreign ministers in December 2001, explored possibilities for giving new impetus and substance to the NATO-Russia relationship.

For more than two decades, NATO has worked to build a partnership with Russia, developing dialogue and practical cooperation in areas of common interest. Cooperation has been suspended since 2014 in response to Russia's military intervention in Ukraine but political and military channels of communication remain open. Concerns about Russia's continued destabilising pattern of military activities and aggressive rhetoric go well beyond Ukraine.

The NATO-Russia Council:

Let us keep remembering that intensive negotiations led to agreement on a joint declaration on "NATO-Russia Relations: A New Quality", signed by Russian and Allied heads of state and government in Rome in May 2002. In this declaration, which builds on the goals and principles of the Founding Act, NATO and Russian leaders pledged to enhance their ability to work together as equals in areas of common interest and to stand together against common threats and risks to their security.

The agreement established the NATO-Russia Council (NRC), which replaced the PJC. The 26 NATO member countries and Russia participate in the NRC as equal partners, identifying and pursuing opportunities for joint decision and joint action across a wide spectrum of security issues in the Euro-Atlantic area. The change from the PJC to the NRC (which met "at 27" rather than in the "NATO+1" format) represented a new philosophical approach to the relationship. This has contributed significantly to creating a strengthened climate of confidence, making political exchange and consultations far more conducive to concrete cooperation.

This way, NRC meetings are chaired by the NATO Secretary General and are held at different levels, at least once a month at the level of ambassadors, twice a year at ministerial level and as needed at summit level. A Preparatory Committee, which meets at least twice a month, supports the work of the NRC and oversees ongoing cooperation. Work in specific areas is developed in the framework of ad hoc or permanent working groups. Meetings are also held once a month between military representatives and twice a year at the level of chiefs of defence staff.

Thus, the NRC and its subordinate structures operate on the principle of consensus and continuous political dialogue. The members of the NRC act in their national capacities and in a manner consistent with their respective collective commitments and obligations. Both NATO members and Russia reserve the right to act independently, although their common objective in the framework of the NRC is to work together in all areas where they have shared interests and concerns. These areas were identified in the Founding Act and cooperation is being intensified on a number of key issues which include the fight against terrorism, crisis management, non-proliferation, arms control and confidence-building measures, theatre missile defence, logistics, military-to-military cooperation, defence reform and civil emergencies. New areas may be added to the NRC's agenda by the mutual consent of its members.

Thereby, the NRC has created several working groups and committees to develop cooperation in these areas and others such as scientific cooperation and challenges of modern society. Views are also exchanged within the NRC on current international issues affecting the security of the Euro-Atlantic area, such as the situations in the Balkans, Afghanistan, Georgia, Ukraine, Iraq, and the broader Middle East region. It also helps build political will for undertaking additional joint practical initiatives, such as an ongoing project to explore its potential contribution to combating the threat posed by Afghan narcotics.

Another concrete manifestation of the level of political cooperation achieved by the NRC, stemming from the frank exchanges of views which it facilitates, is the joint statement adopted by NRC foreign ministers on 9 December 2004, at the height of a serious political crisis taking place in Ukraine. The NRC appealed to all parties to continue to avoid the use or instigation of violence, to refrain from intimidation of voters, and to work to ensure a free and fair electoral process reflecting the will of the Ukrainian people.

The members of the NRC reiterated their support for democracy in Ukraine together with the independence, sovereignty and territorial integrity of the country, elements considered vital for the common and indivisible security and stability of all.

Like so, since its establishment, the NRC has evolved into a productive mechanism for consultation, consensus-building, cooperation, joint decision-making and joint action and has become a fundamental pillar of the NATO-Russia partnership. The positive evolution of NATO-Russia cooperation offers good prospects for its future and its further concrete development. Maintaining this momentum, based on past political and operational achievements, will be an important factor in developing greater opportunities for future cooperation.

Facilitating contacts and cooperation:

Continuing with the previous relations between both parties we will say that a Russian Mission to NATO was established in March 1998 to facilitate NATO-Russia consultation and cooperation. At the beginning of 2001, a NATO Information Office (NIO) was opened in Moscow to improve mutual understanding by disseminating information and publications, and organising conferences and seminars for key target audiences and academies for young students. In particular, the NIO focuses on explaining the rationale for NATO-Russia cooperation, which highlights the increasing number of areas where the interests of NATO member states and Russia converge, such as the fight against terrorism, countering the proliferation of weapons of mass destruction and exchanging experience on defence reform and military transformation.

Likewise, a Military Liaison Mission was established in Moscow in May 2002 to improve transparency and facilitate regular contacts, exchange of information and consultations between NATO's Military Committee and Russia's Ministry of Defence. In 2003, a direct, secure telephone communication link was established between the offices of the NATO Secretary General and the Russian Minister of Defence. A Russian Military Liaison Branch Office was established at NATO's strategic operational command in Mons, Belgium (the Supreme Headquarters Allied Powers in Europe, SHAPE) in 2004. Russian officers are also assigned to the Partnership Coordination Cell at SHAPE to facilitate participation in PfP activities.

The fight against terrorism:

Continuing with the review of the previous relations between the parties, we emphasise that nowhere have positions between NATO member countries and Russia converged more clearly than in the fight against terrorism, which has become a major threat to international security and stability and requires an increasingly coordinated response. In the summer of 2004, a number of tragic events perpetrated by terrorists in Russia brought the NRC together in extraordinary session for the first time in its history. NRC ambassadors strongly condemned terrorism in all its manifestations and renewed their determination to strengthen and intensify common efforts to eliminate this shared threat.

So, concrete steps have followed, including the development of a comprehensive NRC Action Plan on Terrorism approved by NRC foreign ministers on 9 December 2004. The Action Plan gives structure and purpose to NRC cooperation in this key area and consolidates NRC cooperation aimed at preventing terrorism, combating terrorist activities and managing the consequences of terrorist acts.

In that way, joint assessments of specific terrorist threats are being developed and kept under review. Three high-level conferences (in Rome and Moscow in 2002 and in Norfolk, Virginia, United States in 2004) explored the role of the military in combating terrorism, generating recommendations for ways to develop practical military cooperation in this area.

A conference held in Slovenia in June 2005 focused on the challenges encountered by national authorities in Russia, Spain, Turkey and the United States while managing the consequences of recent terrorist acts, and on how lessons learned have since been integrated into policies and practice.

Sand specific aspects of combating terrorism are also a key focus of activities in many areas of cooperation under the NRC, such as civil emergency planning, non-proliferation, airspace management, theatre missile defence, defence reform and scientific cooperation.

Addressing other new security threats:

In this field, the unprecedented threat posed by the increasing availability of ballistic missiles was addressed by cooperation in the area of theatre missile defence, where NATO and Russia achieved impressive results. A groundbreaking joint Command Post Exercise took place in the United States in March 2004, using a computer-simulated scenario to evaluate an experimental concept of operations on theatre missile defence.

This was followed up by a second exercise in the Netherlands in March 2005. The aim is to establish a level of force interoperability that would enable NATO and Russia to work together quickly and effectively to counter ballistic missile threats against NATO and Russian troops engaged in a joint mission. An NRC Interoperability Study in the field of theatre missile defence entered its second phase in 2005.

A groundbreaking joint Command Post Exercise took place in the United States in March 2004, using a computer-simulated scenario to evaluate an experimental concept of operations on theatre missile defence. This was followed up by a second exercise in the Netherlands in March 2005.

The aim is to establish a level of force interoperability that would enable NATO and Russia to work together quickly and effectively to counter ballistic missile threats against NATO and Russian troops engaged in a joint mission. An NRC Interoperability Study in the field of theatre missile defence entered its second phase in 2005.

An NRC initiative on chemical, biological and radiological protection is underway. Joint work is also being taken forward on nuclear issues and cooperation against the proliferation of weapons of mass destruction.

NRC cooperation is also being undertaken in the field of airspace management to enhance air safety and transparency and to counter the threat of the potential use of civilian aircraft for terrorist purposes. Under the Cooperative Airspace Initiative, a feasibility study for reciprocal data exchange is being carried out.

Operational cooperation:

Before 2014, for over seven years, until withdrawing its contingents in summer 2003, Russia provided the largest non-NATO contingents to the NATO-led peacekeeping forces in the Balkans, where Russian soldiers worked alongside Allied and other Partner counterparts to support the international community's efforts to build lasting security and stability in the region. Russian peacekeepers first deployed to Bosnia and Herzegovina in January 1996, where they were part of a multinational brigade in the northern sector, conducting daily patrols and security checks and helping with reconstruction and humanitarian tasks. Having played a vital diplomatic role in securing an end to the Kosovo conflict, despite differences over NATO's 1999 air campaign, Russian troops deployed to Kosovo in June 1999, where they worked as part of multinational brigades in the east, north and south of the province, helped run the Pristina airfield and provided medical facilities and services.

Thus, building on the experience of cooperation in peacekeeping in the Balkans, a generic concept for joint peacekeeping operations is being developed, which would serve as a basis for joint NATO-Russia peacekeeping operations and should provide a detailed scheme of joint work aimed at ensuring smooth, constructive and predictable cooperation between NATO Allies and Russia in case of such an operation.

Modalities were finalised with regard to Russian participation in NATO's maritime Operation Active Endeavour (OAE) in the Mediterranean in December 2004. Russia's contribution to this operation marks the beginning of a new phase of operational activities involving the development of greater military interoperability in relation to both crisis management and the fight against terrorism. NATO member states have also welcomed Russia's offer to provide practical support for the NATO-led International Security Assistance Force in Afghanistan.

On 21 April 2005, in the margins of the informal meeting of NRC foreign ministers in Vilnius, Lithuania, Russia acceded to NATO's PfP Status of Forces Agreement, thereby establishing a necessary legal framework for further operational activities and intensified practical cooperation.

Defence reform:

Russia and NATO member countries share an interest in defence reform, given the common need for armed forces that are appropriately sized, trained and equipped to deal with the full spectrum of 21st century threats. NRC cooperation in the area of defence reform covers many different aspects including resource management, defence industry conversion, defence and force planning, and macro-economic, financial and social issues.

Other areas of cooperation include managing military nuclear waste, strategic air transport and military infrastructure engineering, and logistics interoperability, a prerequisite for effective cooperation across the board. Exploratory work on how to improve the general interoperability of NATO and Russian forces is also underway. Moreover, two fellowships for Russian scholars have been set up at the NATO Defence College in Rome to promote research on defence reform.

In June 2001 a NATO-Russia Retraining Centre was established in Moscow for discharged military personnel and their families, and in 2003 a further six regional retraining centres spread throughout Russia were set up.

Military-to-military cooperation:

A main objective of military-to-military cooperation is to improve inter-operability between Russian and Allied forces, since modern militaries must be able to operate within multinational command and force structures when called upon to work together in peace-support or other crisis-management operations. A substantial exercise and training programme is being implemented.

In the wake of the loss of the Russian nuclear submarine Kursk, in August 2000, joint work has also been undertaken on submarine crew escape and rescue. A Framework Agreement was signed in February 2003, which represented an important step towards standardising search and rescue procedures, collaborating in equipment development, exchanging information and facilitating joint exercises to test procedures.

Besides, a framework for reciprocal naval exchanges and port visits is being developed, and possible activities to enhance exercises between NATO and Russian naval formations are being developed. In 2005, training exercises were conducted to help prepare Russian crews for future support for Operation Active Endeavour.

Logistics cooperation:

Let us be aware that today's security environment calls for more mobile forces and multinational operations, which require improved coordination and the pooling of resources wherever possible. Various NRC initiatives are pursuing cooperation in the area of logistics on both the civilian and military side. Meetings and seminars have focused on establishing a sound foundation of mutual understanding in this field by promoting information-sharing in areas such as logistics policies, doctrine, structures and lessons learned.

So, opportunities for practical cooperation are being explored in areas such as air transport and air-to-air refuelling. Such practical cooperation between NATO and Russia will be significantly facilitated by the PfP Status of Forces Agreement, once ratified, as well as by a Transit Agreement and a memorandum of understanding on Host Nation Support which were being finalised in 2005.

Progress in arms control:

In the period prior to 2014, NATO and Russia also discuss issues related to arms control and confidence-building measures. Within this framework they have reaffirmed their commitment to the Treaty on Conventional Forces in Europe (CFE) as a cornerstone of European security. Progress in this sphere is dependent upon Russia's implementation of its remaining commitments articulated in the Final Act of the 1999 Istanbul Conference of the State Parties to the CFE Treaty, with respect to Georgia and Moldova.

NATO member states have stated that fulfilment of these commitments will create the necessary conditions for achieving ratification of the Agreement on Adaptation of the

CFE Treaty by all 30 States Parties to the Treaty and securing its entry into force.

Civil emergencies and disaster relief:

Equally, cooperation between NATO and Russia in the area of disaster response dates back to the signing of a memorandum of understanding on civil emergency planning and disaster preparedness in 1996. Practical forms of cooperation include joint work to better prepare for protecting civilian populations and responding to different kinds of emergency situation. A Russian proposal led to the establishment at NATO in 1998 of a Euro-Atlantic Disaster Response Coordination Centre (EADRCC), which was used to coordinate assistance from EAPC member countries to refugee relief operations during the Kosovo conflict and has subsequently been called upon to coordinate relief in the wake of flooding, earthquakes, landslides, fires and other disasters in different Partner countries.

Likewise, in the framework of the NRC, work has initially concentrated on improving interoperability, procedures and the exchange of information and experience. Russia hosted civil emergency planning and response exercises in 2002 and 2004. Russia has co-sponsored an initiative with Hungary to develop a rapid response capability in the event of an emergency involving chemical, biological, radiological or nuclear agents.

Scientific and environmental cooperation:

Scientific cooperation with Russia dates back to 1998, when a Memorandum of Understanding on Scientific and Technological Cooperation was signed. More scientists from Russia than from any other Partner country have benefited from fellowships and grants under NATO's science programmes. A key focus of current scientific cooperative activities under the NRC is the application of civil science to defence against terrorism and new threats, such as in explosives detection, examining the social and psychological impact of terrorism, protection against chemical, biological, radiological or nuclear agents, cyber-security and transport security. Another area of collaboration is the forecasting and prevention of catastrophes. Environmental protection problems arising from civilian and military activities are another important area of cooperation.

Relations between NATO and Russia after 2014:

As we have previously said, NATO worked for more than twenty years to build a partnership with Russia, developing dialogue and practical cooperation in areas of common interest. But this cooperation was suspended since 2014 in response to Russia's military intervention in Ukraine, but the political and military channels of communication remain open. Currently, such concerns about Russia's military activities and aggressive rhetoric go far beyond Ukraine.

After Russia's illegal and illegitimate annexation of Crimea in March 2014, the Allies suspended all practical civilian and military cooperation in April 2014, while keeping open channels of political and military communication. The NATO-Russia Council (NRC) remains an important forum for dialogue, on the basis of reciprocity, and has met ten times since 2016.

At the NATO Summit in Wales in September 2014 and at successive summits since then, Allied leaders have condemned in the strongest terms Russia's military intervention in Ukraine, calling on Russia to stop and withdraw its forces from Ukraine and along the country's border.

Allies continue to demand that Russia comply with international law and its international obligations and responsibilities; end its illegitimate occupation of Crimea; refrain from aggressive actions against Ukraine; halt the flow of weapons, equipment, people and money across the border to the separatists; and stop fomenting tension along and across the Ukrainian border. NATO does not and will not recognise Russia's illegal and illegitimate annexation of Crimea.

The Allies have also noted that violence and insecurity in the region led to the tragic downing of Malaysia Airlines passenger flight MH17 on 17 July 2014, calling for those directly and indirectly responsible to be held accountable and brought to justice as soon as possible. In May 2018, the Joint Investigation Team, which is investigating the MH17 crash, concluded that the BUK-TELAR that was used to down the aircraft originated from the 53rd Anti-Aircraft Missile Brigade, a unit of the Russian army from Kursk. Allies stand in solidarity with the Netherlands and Australia, which call on Russia to take State responsibility for the downing of flight MH17.

Allies strongly support the settlement of the conflict in eastern Ukraine by diplomatic and peaceful means and welcome the ongoing diplomatic efforts to this end. All signatories of the Minsk Agreements must comply with their commitments and ensure their full implementation. Russia has a significant responsibility in this regard.

NATO's concerns go well beyond Russia's activities in Ukraine. The Allies continue to express their support for the territorial integrity of Georgia and the Republic of Moldova within their internationally recognised borders and call on Russia to withdraw the forces it has stationed in all three countries without their consent. Russia's military activities, particularly along NATO's borders, have increased and its behavior continues to make the Euro-Atlantic security environment less stable and predictable, in particular its practice of calling snap exercises, deploying near NATO borders, conducting large-scale training and exercises and violating Allied airspace. Russia is also challenging Euro-Atlantic security and stability through hybrid actions, including attempted interference in the election processes and the sovereignty of nations, widespread disinformation campaigns and malicious cyber activities. The Allies also condemn the use of a military-grade nerve agent in Salisburyin March 2018, and express solidarity with the United Kingdom. In the wake of this attack, the maximum number of personnel in the Russian Mission at NATO Headquarters was reduced by 10 people.

This is compounded by Russia's continued violation, non-implementation and circumvention of numerous obligations and commitments in the realm of arms control and confidence- and security-building measures. Allies have long been concerned about Russia's ongoing selective implementation of the Vienna Document and the Open Skies Treaty, and its long-standing non-implementation of the Conventional Forces in Europe Treaty which undermine Euro-Atlantic security. Moreover, in December 2018,

NATO foreign ministers supported the finding of the United States that Russia was in material breach of its obligations under the Intermediate-Range Nuclear Forces (INF) Treaty not to possess, produce or flight-test a ground-launched cruise missile with a range capability of 500 to 5,500 kilometers, or to possess or produce launchers of such missiles. The Allies concluded that Russia had developed and fielded a missile system, the SSC-8 (9M729), which violated the Treaty and posed significant risks to Euro-Atlantic security, and called on Russia to return urgently to full and verifiable compliance.

On 1 February 2019, the United States suspended its obligations under the INF Treaty, providing the requisite six-month written notice to Treaty Parties of its withdrawal. The Allies remained open to dialogue and engaged Russia on its violation, including at two NATO-Russia Council meetings in January and July 2019.

However, Russia continued to deny its INF Treaty violation, refused to provide any credible response, and took no demonstrable steps toward returning to full and verifiable compliance. As a result, on 2 August, the United States decided to withdraw from Treaty with the full support of the Allies. NATO will respond in a measured and responsible way to the significant risks posed by Russia's SSC-8 system. At the same time, Allies are firmly committed to the preservation of effective international arms control, disarmament and non-proliferation. For over three decades, the INF Treaty was a landmark in arms control. It entered into force in 1988 with the aim to reduce threats to security and stability in Europe, in particular the threat of short-warning attack on targets of strategic importance, by requiring the verifiable elimination of an entire class of missiles possessed by the United States and the former Soviet Union.

Russia's military intervention and considerable military presence in Syria have posed further risks for the Alliance. On 5 October 2015, in response to Russia's military intervention in Syria, the Allies called on Russia to immediately cease their attacks on the Syrian opposition and civilians, to focus its efforts on fighting so-called Islamic State, and to promote a solution to the conflict through a political transition. In April 2018, Allies expressed strong support to the US, UK and French joint military action in response to the use of chemical weapons in Syria.

As mentioned above, for more than two decades, NATO has worked to build a partnership with Russia, including through the mechanism of the NRC, based upon the 1997 NATO-Russia Founding Act and the 2002 Rome Declaration. Russia has breached its commitments, as well as violated international law, breaking the trust at the core of its cooperation with NATO.

The Allies continue to believe that a partnership between NATO and Russia, based on respect for international law, would be of strategic value. They continue to aspire to a cooperative, constructive relationship with Russia (including reciprocal confidence-building and transparency measures and increased mutual understanding of NATO's and Russia's non-strategic nuclear force postures in Europe) based on common security concerns and interests, in a Europe where each country freely chooses its future. They regret that the conditions for that relationship do not currently exist. Meeting at the Brussels Summit in July 2018, Allied leaders underlined that there can be no return to "business as usual" until there is a clear, constructive change in Russia's actions that demonstrates compliance with international law and its international obligations and responsibilities.


**NATO and Ukraine**

As we know, NATO's relationship with Ukraine has developed progressively since the country gained independence in 1991. Given Ukraine's strategic position as a bridge between eastern and western Europe, NATO-Ukraine relations are central to building peace and stability within the Euro-Atlantic region.

NATO and Ukraine are actively engaged in international peace-support operations and in addressing common security challenges. Over the years, a pattern of dialogue and practical cooperation in a wide range of other areas has become well established. A key aspect of the partnership is the support given by NATO and individual member countries for Ukraine's reform efforts, which received renewed momentum following the dramatic events of the 2004 "Orange Revolution" and remain critical to Ukraine's aspirations to closer Euro-Atlantic integration.

A sovereign, independent and stable Ukraine, firmly committed to democracy and the rule of law, is key to Euro-Atlantic security. Relations between NATO and Ukraine date back to the early 1990s and have since developed into one of the most substantial of NATO's partnerships. Since 2014, in the wake of the Russia-Ukraine conflict, cooperation has been intensified in critical areas.

The evolution of relations:

In 1991, formal relations between NATO and Ukraine began when Ukraine joined the North Atlantic Cooperation Council (later replaced by the Euro-Atlantic Partnership Council) immediately upon achieving independence following the break-up of the Soviet Union. In 1994, Ukraine became the first member state of the Commonwealth of Independent States (CIS) to join the Partnership for Peace (PfP). During the 1990s, the country also demonstrated its commitment to contributing to Euro-Atlantic security through its support for NATO-led peacekeeping operations in the Balkans.

Later, in Madrid on 9 July 1997, the Ukrainian president and NATO heads of state and government signed a Charter for a Distinctive Partnership between NATO and Ukraine. It provides the formal basis for NATO-Ukraine relations and was an opportunity for NATO member countries to reaffirm their support for Ukrainian sovereignty and independence, territorial integrity, democratic development, economic prosperity and status as a non-nuclear weapons state, as well as for the principle of inviolability of frontiers. The Alliance regards these as key factors of stability and security in Central and Eastern Europe and on the continent as a whole.

In addition, the Charter also established the NATO-Ukraine Commission (NUC), which is the decision-making body responsible for developing the relationship between NATO and Ukraine and for directing cooperative activities. It provides a forum for consultation on security issues of common concern and is tasked with ensuring the proper implementation of the Charter's provisions, assessing the overall development of the NATO-Ukraine relationship, surveying planning for future activities and suggesting ways of improving or further developing cooperation. It is also responsible for reviewing cooperative activities organised within different frameworks such as the Partnership for Peace, as well as activities in the military-to-military sphere, developed in the context of Annual Work Plans undertaken under the auspices of the Military Committee with Ukraine.

All NATO member countries and Ukraine are represented in the NUC, which meets regularly at the level of ambassadors and military representatives, and periodically at the level of foreign and defence ministers and chiefs of staff, as well as at summit level. Joint working groups have been set up to take work forward in specific areas, namely defence reform, armaments, economic security, and scientific and environmental cooperation, which are areas identified by the Charter for political consultation and practical cooperation. Other areas include operational issues, crisis management and peace support, military-tomilitary cooperation and civil emergency planning.

Shortly before the fifth anniversary of the signing of the Charter in May 2002, then-President Leonid Kuchma announced Ukraine's goal of eventual NATO membership. Later that month, at the meeting of the NUC at ministerial level in Reykjavik, foreign ministers underlined their desire to take their relationship forward to a qualitatively new level, including through intensified consultations and cooperation on political, economic and defence issues.

The NATO-Ukraine Action Plan:

In November 2002, the NATO-Ukraine Action Plan was adopted at a NUC meeting of foreign ministers in Prague.

The Action Plan is built on the Charter, which remains the basic foundation underpinning NATO-Ukraine relations. Its purpose is to clearly identify Ukraine's strategic objectives and priorities in pursuit of its aspirations for full integration into Euro-Atlantic security structures and to provide a strategic framework for existing and future cooperation. It also aims to deepen and broaden the NATO-Ukraine relationship and sets out jointly agreed principles and objectives covering political and economic issues, information issues, security, defence and military issues, information protection and security, and legal issues.

Thus, the adoption of the Action Plan (at a time when the Alliance expressed grave concerns about reports of the authorisation at the highest level of the transfer of air-defence equipment from Ukraine to Iraq) demonstrated the strength of the Allies' commitment to develop strong NATO-Ukraine relations and to encourage Ukraine to work towards closer Euro-Atlantic integration. NATO countries urged Ukraine to take the reform process forward vigorously in order to strengthen democracy, the rule of law, human rights and the market economy. They also emphasised that helping Ukraine to transform its defence and security sector institutions is a key priority of NATO-Ukraine cooperation.

So, to support the implementation of the Action Plan's objectives, Annual Target Plans are agreed within the framework of the NATO-Ukraine Commission. Ukraine sets its own targets in terms of the activities it wishes to pursue both internally and in cooperation with NATO. The NUC monitors their implementation; assessment meetings are held twice a year while a progress report is prepared annually.

Taking stock of progress under the Action Plan in mid-2004, the Allies emphasised the need for Ukraine's leadership to take firm steps to ensure a free and fair electoral process, guaranteed media freedoms and rule of law, strengthened civil society and judiciary, improved arms export controls and progress on defence and security sector reform and the allocation of financial support to its implementation. They also acknowledged the substantial progress that had been made in pursuing defence reform in Ukraine in 2004, in particular the completion by Ukraine of a comprehensive defence review, which called for a major overhaul of Ukraine's defence posture to be coupled with major modernisation of the Ukrainian armed forces.

The Allies closely followed political developments surrounding the presidential elections in Ukraine in the autumn of 2004, where the legitimacy of the results of the second round was contested by the opposition and by international observers, leading to the Orange Revolution of popular protest and a courtordered re-run of the second round. Under these circumstances, it was decided to postpone a meeting of the NATO-Ukraine Commission at ministerial level, scheduled for December 2004, until a later date. In announcing this decision, NATO's Secretary General emphasised that respect for free and fair elections constituted one of the basic principles underlying the Distinctive Partnership.

Viktor Yushchenko won the re-run of the second round. Shortly after his inauguration in January 2005, he was invited to a summit meeting in Brussels on 22 February. NATO leaders expressed support for the new President's ambitious reform plans for Ukraine, which corresponded broadly to the objectives undertaken by Ukraine in the

NATO-Ukraine Action Plan. They agreed to sharpen and refocus NATO Ukraine cooperation in line with the new government's priorities.

The Intensified Dialogue:

In that way, two months later, at the NUC meeting of foreign ministers in Vilnius, Lithuania, on 21 April 2005, the Allies and Ukraine launched an Intensified Dialogue on Ukraine's aspirations to NATO membership. The first concrete step in this process was taken on 27 June 2005, during a visit by the Secretary General to Kyiv, when the Ukrainian government formally presented an initial discussion paper.

The discussion paper addressed key issues set out in the 1995 Study on NATO Enlargement (domestic and foreign policy, defence and security sector reform as well as legal and security issues) and highlighted in specific terms those areas where progress would be needed to bring Ukraine's aspirations closer to reality. This paper has provided the basis for the holding of structured expert discussions, launched in September 2005, which give Ukrainian officials the opportunity to learn more about what would be expected from Ukraine as a potential member of the Alliance and also allow NATO officials to examine Ukrainian reform policy and capabilities in greater detail.

And at Vilnius, the Allies and Ukraine also announced a package of short-term actions designed to enhance NATO-Ukraine cooperation in key reform areas: strengthening democratic institutions, enhancing political dialogue, intensifying defence and security sector reform, improving public information, and managing the social and economic consequences of reform. These are high priorities for the Ukrainian government, as they are vital to the success of the democratic transformation that the Ukrainian people demanded in December 2004. These are also areas where NATO can offer specific expertise and, in some cases, material assistance.

Besides, the Intensified Dialogue addresses issues specifically related to Ukraine's possible NATO membership. The package of short-term measures is designed to focus practical cooperation in support of urgent reform goals. Both of these initiatives are intended to complement and reinforce existing cooperation in the framework of the NATO-Ukraine Action Plan.

Therefore, the launch of the Intensified Dialogue with Ukraine marks a real milestone in NATO-Ukraine relations and in Ukraine's pursuit of Euro-Atlantic integration. It is a clear signal from NATO Allies that they support Ukraine's aspirations.

Nonetheless, this process does not guarantee an invitation to join the Alliance, such an invitation would be based on Ukraine's performance in the implementation of key reform goals. NATO and individual Allies are committed to providing assistance and advice, but the pace of progress remains in Ukraine's hands.

Facilitating contacts and cooperation:

Remember that Ukraine was one of the first countries to open a diplomatic mission to NATO in 1997, and a Military Liaison Mission was opened in 1998. Ukrainian military personnel also serve at the Partnership Coordination Cell located at NATO's military operational headquarters in Mons, Belgium.

NATO opened an Information and Documentation Centre (NIDC) in Kyiv in May 1997. The role of the Centre is to provide a focal point for information activities designed to promote the mutual benefits of Ukraine's partnership with NATO and explain Alliance policies to the Ukrainian public. The Centre seeks to disseminate information and stimulate debate on Euro-Atlantic integration and security issues through publications, seminars, conferences and information academies for young students and civil servants. Moreover, the Centre has recently opened a series of information points in several regions of the country outside Kyiv.

A civilian-led NATO Liaison Office (NLO) was established in Kyiv in April 1999 to work directly with Ukrainian officials to encourage them to make full use of opportunities for cooperation under the NATO-Ukraine Charter and the PfP programme. It is active in supporting Ukraine's efforts to reform its defence and security sector, in strengthening cooperation under the Action Plan, and in facilitating contacts between NATO and Ukrainian authorities at all levels.

The Office also has a military liaison element that works closely with Ukraine's armed forces to facilitate participation in joint training, exercises, and NATO-led peacekeeping operations. In August 2004, a NATO-Ukraine Defence Documentation Office was also opened to improve access to documentation for units and staffs of the armed forces involved in PfP activities.

Peace-support and security cooperation:

On the other hand, Ukraine has over the years contributed an infantry battalion, a mechanised infantry battalion and a helicopter squadron to the NATO-led peacekeeping force in Bosnia and Herzegovina. Deployments to the NATO-led operation in Kosovo have included a helicopter squadron as well as nearly 300 peacekeepers, who continue to serve in the US-led sector as part of the joint Polish-Ukrainian battalion.

Thus, Ukraine is further contributing to international stability and the fight against terrorism by providing overflight clearance for forces deployed in Afghanistan as part of the NATO-led International Security Assistance Force (ISAF), or as part of the coalition forces under the US-led Operation Enduring Freedom. Some 1600 Ukrainian troops were also deployed to Iraq, as part of a Polish-led multinational force in one of the sectors of the international stabilisation force, which includes peacekeepers from several NATO and Partner countries.

Finally, the Allies have welcomed Ukraine's offer to support Operation Active Endeavour, NATO's maritime operation in the Mediterranean aimed at helping deter, disrupt and protect against terrorism. An exchange of letters signed by NATO Secretary General Jaap de Hoop Scheffer and Minister of Foreign Affairs Borys Tarasyuk in Vilnius on 21 April 2005 set out agreed procedures for Ukraine's support. This paved the way for contacts at the working level to discuss plans to integrate Ukraine's contribution into the operation. Ukraine's support will further enhance NATO-Ukraine cooperation against terrorism and interoperability between NATO and Ukrainian military forces and add another concrete dimension to Ukraine's already impressive array of contributions to Euro-Atlantic security.

Defence and security sector reform:

We want to highlight that cooperation in the area of defence and security sector reform has been crucial to the ongoing transformation of Ukraine's security posture and remains an essential part of its democratic transition. Since gaining independence in 1991 and inheriting a significant part of the armed forces of the former Soviet Union, Ukraine has been in the process of establishing and then reforming its armed forces and other parts of its security structures in order to bring them in line with the requirements of the changed security environment, democratic conditions and available resources.

It has sought NATO's support in helping to transform massive conscript forces into smaller, professional, more mobile armed forces capable of meeting its security needs as well as contributing actively to European stability and security. Priorities for NATO in this context are the strengthening of the democratic and civilian control of Ukraine's armed forces and improving their interoperability with NATO forces.

So, recognising the importance of this process, NATO has extended practical assistance in managing defence and security sector reforms. The NATO-Ukraine Joint Working Group on Defence Reform (JWGDR) is the primary focus for cooperation in defence and security sector reform. It was established in 1998, under the auspices of the NATO-Ukraine Commission, to pursue initiatives in the area of civil-military relations, democratic control of the armed forces, defence planning, policy, strategy and national security concepts. All NATO member countries and Ukraine are represented in annual meetings of the JWGDR at senior level, co-chaired by NATO's Assistant Secretary General for Defence Policy and Planning and the Deputy Secretary of Ukraine's National Security and Defence Council and bringing together high-ranking officials from NATO member states and Ukraine. Once a year, the JWGDR organises high-level informal consultations on defence reform and defence policy involving Ukrainian and NATO Defence Ministers, as well as key defence and security experts.

In this sense, the JWGDR allows Ukraine to draw on the experience and expertise of NATO countries and serves as a channel for providing assistance. It also provides the institutional basis for cooperation with ministries and agencies engaged in supporting defence and security sector reform in Ukraine. Key aspects of cooperation have included helping Ukraine to develop a new security concept and military doctrine, as well as providing support for defence budgeting and planning, military downsizing and conversion, the establishment of rapid reaction forces, professionalisation of the armed forces and the completion and implementation of a comprehensive defence review.

In addition, the JWGDR has launched several initiatives aimed at supporting the transformation of the Ukrainian security posture, including the use of the PfP Planning and Review Process (PARP) to support reforms in individual components of Ukraine's security sector, the provision of assistance in managing the process of the defence review, the organisation of roundtables with the Ukrainian

Parliament (the Verkhovna Rada), with the participation of NATO experts and representatives of the NATO Parliamentary Assembly, on topics linked to the legislative programme of the Ukrainian Parliament, support for efforts to strengthen the role of civilians in the Ministry of Defence and establish an effective defence organisation, the arrangement of meetings to harmonise bilateral assistance to Ukraine, and various forms of cooperation with the Ukrainian Border Guard, the Ministry of the Interior and the Ministry of Emergencies. Activities are thus not limited to the armed forces or the Ministry of Defence but aim to provide support for reforms undertaken in all the security sector institutions.

It is interesting to mention that, Ukraine's drive to reform its defence and security sector also benefits from participation in the Partnership for Peace and in the PfP Planning and Review Process, which enables joint goals to be developed for shaping force structures and capabilities to help Ukraine to meet its objectives for interoperability with the Alliance. Ukraine's participation in other PfP activities, through its annual Individual Partnership Programme, has also remained steady and, in 2005, is expected to include participation in some 400 activities, including language training, military exercises and consideration of operational concepts.

Equally, much-needed assistance in implementing demilitarisation projects is being channelled through PfP Trust Funds, which permit individual NATO countries to pool voluntary financial contributions so as to increase their collective impact on the demilitarisation process. The first such project, implemented by the NATO Maintenance and Supply Agency (NAMSA) as the executing agency, resulted in the safe destruction of 400 000 landmines and was the first step in destroying Ukraine's stockpile of almost seven million anti-personnel mines. Canada was the lead country for the project, supported by financial contributions from Hungary, Poland and the Netherlands. A second PfP Trust Fund project to destroy 133 000 tons of conventional munitions, 1.5 million small arms and other weapons was launched in 2005. The first three-year phase of the project, which is to be carried out over an estimated twelve years, will be led by the United States. It will be the largest demilitarisation project of its kind ever undertaken.

Economic aspects of defence:

From another point of view, cooperation with Ukraine on economic aspects of defence has two main axes: retraining activities, and institutional dialogue, concentrating on issues related to defence economics, economic security and economic aspects of defence industry restructuring and Euro-Atlantic integration.

Moreover, exchanges of experience are promoted with experts on security aspects of economic developments, including defence budgets and management of defence resources and their relationship with the macro-economy and restructuring in the defence sector. Courses in defence economics (covering the whole budgetary process from financial planning to financial control) have also been organised.

Equally, managing the economic and social consequences of defence reform is a key area of cooperation. Under an agreement with Ukraine's National Coordination Centre, which is in charge of social adaptation of discharged military servicemen, NATO is financing and implementing language and management courses in Ukraine. NATO has doubled the resources devoted to cooperation in this area, following the launch of the Intensified Dialogue with Ukraine.

Military-to-military cooperation:

In this area, cooperation between NATO and Ukrainian militaries is developed in the framework of the NATO-Ukraine Military Work Plan under the auspices of the Military Committee with Ukraine. A key focus is to help Ukraine implement its defence reform objectives, complementing the work carried out under the JWGDR with military expertise.

So, NATO military staff have also taken the lead in developing a legal framework to enable NATO and Ukraine to further develop operational cooperation. These include the PfP Status of Forces Agreement (SOFA) and its additional protocol, which was ratified on 1 March 2000 by the Ukrainian Parliament and entered into force on 26 May 2000. This agreement exempts participants in PfP events from passport and visa regulations and immigration inspection on entering or leaving the territory of a receiving state and thereby facilitates Ukrainian participation in PfP military exercises.

A memorandum of understanding on Host Nation Support, ratified in March 2004, addresses issues related to the provision of civil and military assistance to Allied forces located on, or in transit through, Ukrainian territory in peacetime, crisis or war. A Memorandum of Understanding on Strategic Airlift was signed in June 2004, which will permit Ukraine to make a substantial contribution to NATO's capability to move outsized cargo.

Thus, a wide range of PfP activities and military exercises, sometimes hosted by Ukraine, allow military personnel to train for peace-support operations and gain hands-on experience of working with forces from NATO countries and other partners.

Senior Ukrainian officers regularly participate in courses at the NATO Defence College in Rome, Italy, and the NATO School at Oberammergau, Germany. Contacts with these establishments have been instrumental in setting up a new multinational faculty at the Ukrainian Defence Academy.

Armaments cooperation:

In this regard, technical cooperation between Ukraine and NATO in the field of armaments focuses on enhancing interoperability between defence systems to facilitate Ukrainian contributions to joint peace-support operations. Cooperation in this area started when Ukraine joined the PfP programme and began participating in an increasing number of the armaments groups which meet under the auspices of the Conference of National Armaments Directors (CNAD), a senior NATO body which identifies opportunities for cooperation between countries in defence equipment procurement processes, focusing in particular on technical standards.

A Joint Working Group on Armaments, which met for the first time in March 2004, is supporting the further development of cooperation in this area.

Civil emergency planning:

In Ukraine, western regions are prone to heavy flooding, and NATO countries and other partners provided assistance after severe floods in 1995, 1998 and 2001. Since 1997, in accordance with a memorandum of understanding on Civil Emergency Planning and Disaster Preparedness, a key focus of cooperation has been to help Ukraine to prepare better for such emergencies and to manage their consequences more effectively. PfP exercises, including one held in Ukraine's Trans-Carpathian region in September 2000, help to test disaster-relief procedures. A project was launched in 2001, involving neighbouring countries, to develop an effective flood-warning and response system for the Tisza River catchment area. PfP exercises also help develop plans and effective disaster-response capabilities to deal with other natural emergencies such as avalanches and earthquakes, or man-made accidents or terrorist attacks involving toxic spills or chemical, biological, radiological or nuclear agents.

Science and environmental issues:

In 1991, Ukraine's participation in NATO science programmes began and has since been boosted by the creation of a Joint Working Group on Scientific and Environmental Cooperation. Over the years, Ukraine has been second only to Russia in terms of benefiting from NATO grants for scientific collaboration. In addition to applying science to defence against terrorism and new threats, in line with the new direction of NATO's science programme, Ukraine's priority areas for cooperation include information technologies, cell biology and biotechnology, new materials, environmental protection, and the rational use of natural resources. Environmental cooperation focuses in particular on defencerelated environmental problems. NATO has also sponsored several projects to provide basic infrastructure for computer networking among Ukrainian research communities and to facilitate their access to the Internet.

Public information:

On the other hand, as the Intensified Dialogue process moves forward, it will be important for the Ukrainian administration to convince the Ukrainian people that its ambitious reform programme and its aspirations to NATO membership are in the country's interest. It is clear that many people in Ukraine are still suspicious of NATO and associate the Alliance with Cold War stereotypes.

The Allies have offered, as part of the short-term actions agreed at Vilnius, to cooperate with the Ukrainian authorities in raising awareness about what NATO is today and in better explaining the NATO-Ukraine relationship.

Encouraging people to take a fresh look at the Alliance would allow them to discover how NATO has transformed itself since the end of the Cold War and has developed new partnerships throughout the Euro-Atlantic area to meet new security challenges, including a strategic relationship with Russia.

The Ukrainian public also needs to be made more aware of the pattern of dialogue and practical, mutually beneficial cooperation between NATO and Ukraine, which has become well established in a wide range of areas over the past decade. This shared experience of cooperation will provide a solid foundation for the further deepening of the NATO-Ukraine relationship in the years to come.

Relations with Ukraine today:

A sovereign, independent and stable Ukraine, firmly committed to democracy and the rule of law, is key to Euro-Atlantic security. Relations between NATO and Ukraine date back to the early 1990s and have since developed into one of the most substantial of NATO's partnerships. Since 2014, in the wake of the Russia-Ukraine conflict, cooperation has been intensified in critical areas.

From the very beginning of the Russia-Ukraine conflict, NATO has adopted a firm position in full support of Ukraine's sovereignty and territorial integrity within its internationally recognised borders. The Allies immediately condemned - and have since then repeatedly stated that they will not recognise - Russia's illegal and illegitimate "annexation" of Crimea in March 2014. They also condemned Russia's deliberate destabilisation of eastern Ukraine caused by its military intervention and support for the militants. The Allies decided to suspend all practical civilian and military cooperation with Russia, while leaving political and military channels of communication open. Since then, Allied Ambassadors reiterate NATO's firm position on Ukraine's territorial integrity and sovereignty at meetings of the NATO-Russia Council, which continues to meet periodically.

Throughout the crisis, regular consultations have taken place in the NATO-Ukraine Commission (NUC) in view of the direct threats faced by Ukraine to its territorial integrity, political independence and security. Allied leaders met with

President Petro Poroshenko at the NATO summits in Wales (2014), Warsaw (2016) and Brussels (2018). Foreign and defence ministers as well as ambassadors regularly discussed the security situation in and around Ukraine. Joint statements issued by NUC foreign ministers in April 2014, December 2014 and May 2015 and by Heads of State and Government at the NATO summit meetings in Wales and Warsaw demonstrate NATO's unwavering support for and solidarity with Ukraine.

The Allies have also pledged to support the efforts of the Ukrainian government to implement wide-ranging reforms to meet the aspirations of Ukrainian people to see their country firmly anchored among European democracies.

In parallel to its political support to Ukraine, NATO has significantly stepped up its practical assistance to Ukraine. Immediately following the illegal and illegitimate "annexation" of Crimea by Russia, NATO foreign ministers agreed on measures to enhance Ukraine's ability to provide for its own security. They also decided to further develop their practical support to Ukraine, based on a significant enhancement of existing cooperation programs as well as the development of substantial new programs. At the summit in Warsaw, NATO's measures in support of Ukraine became part of the Comprehensive Assistance Package (CAP). The CAP is designed to support Ukraine's ability to provide for its own security and to implement wide-ranging reforms, including as set out in Ukraine's Strategic Defence Bulletin of 2016. It comprises eight Trust Funds set up exclusively for Ukraine, working in critical areas of reform and capability development in Ukraine's security and defence sector.

The 1997 Charter on a Distinctive Partnership remains the basic foundation underpinning NATO-Ukraine relations. The NATO-Ukraine Commission (NUC) directs cooperative activities and provides a forum for consultation between the Allies and Ukraine on security issues of common concern. The NUC can meet at various levels, including heads of state and government, ministers of foreign affairs or defence, ambassadors and in various working-level formats.

The Declaration to Complement the Charter, signed in 2009, gave the NUC a central role in deepening political dialogue and cooperation to underpin Ukraine's reform efforts. The principal tool to support this process is the Annual National Program (ANP), which reflects Ukraine's national reform objectives and annual implementation plans. The ANP is composed of five chapters focusing on: political and economic issues; defence and military issues; resources; security issues; and legal issues.

Allies assess progress under the ANP annually. The responsibility for implementation falls primarily on Ukraine and is coordinated by the office of the Vice Prime Minister for European and Euro-Atlantic Integration, who also chairs Ukraine's Commission for Cooperation with NATO. Through the ANP process, Allies encourage Ukraine to take the reform process forward vigorously to strengthen democracy, the rule of law, human rights and the market economy. Helping Ukraine achieve a far-reaching transformation of the defence and security sector is another priority.

Joint working groups have been set up under the auspices of the NUC, to take work forward in specific areas. They include the Joint Working Group on Defence Reform

(JWGDR), the Joint Working Group on Defence Technical Cooperation (JWGDTC), the Joint Working Group on Scientific and Environmental Cooperation (JWGSEC) and the Joint Working Group on Civil Emergency Planning (JWGCEP).

The NATO Representation to Ukraine supports cooperation on the ground. It consists of the NATO Information and Documentation Center, established in 1997 to support efforts to inform the public about NATO's activities and the benefits of NATO-Ukraine cooperation, and the NATO Liaison Office, established in 1999 to facilitate Ukraine's participation in NATO's Partnership for Peace program and to support its defence and security sector reform efforts by liaising with the foreign ministry, defence ministry, National Security and Defence Council, and other Ukrainian agencies. The NATO Representation to Ukraine leads on the provision of strategic-level advice under NATO's Comprehensive Assistance Package for Ukraine.

Key areas of cooperation today:

Consultations and cooperation between NATO and Ukraine cover a wide range of areas including peace-support operations, defence and security sector reform, military-to-military cooperation, defence technology, interoperability and industry, civil preparedness, science and environment, and public diplomacy. Cooperation in many areas is being intensified to enhance Ukraine's ability to provide for its own security in the wake of the conflict with Russia and its efforts to implement wide-ranging reforms.

- Peace-support operations

Ukraine has long been an active contributor to Euro-Atlantic security by deploying troops that work with peacekeepers from NATO and other partner countries. In spite of the Russia-Ukraine conflict, Ukraine continues to contribute to NATO-led operations and missions.

Ukraine has supported NATO-led peace-support operations in the Balkans - both Bosnia and Herzegovina, and Kosovo. It continues to contribute to the Kosovo Force (KFOR), currently with a heavy engineering unit with counter-improvised explosive devices capabilities.

In support of the NATO-led International Security Assistance Force (ISAF) in Afghanistan, Ukraine allowed for over-flight clearance and the transit of supplies for forces deployed there. Ukraine also contributed medical personnel to support Provincial Reconstruction Teams in Afghanistan and instructors to the NATO Training Mission in Afghanistan. Following the completion of ISAF's mission at the end of 2014, Ukraine is currently supporting the NATO-led mission to train, advise and assist Afghan security forces, known as the Resolute Support mission.

From March 2005, Ukraine contributed officers to the NATO Training Mission in Iraq, which terminated in December 2011.

Ukraine has deployed ships in support of Operation Active Endeavor - NATO's maritime operation in the Mediterranean aiming to helping deter, disrupt and protect against terrorism - six times since 2007, most recently in November 2010. At the end of 2013, it also contributed to frigate to NATO's Operation Ocean Shield, which fought piracy off the coast of Somalia. Since the creation of maritime operation Sea Guardian in 2016, Ukraine continues to provide information in support of NATO's maritime situational awareness in and around the Black Sea.

Ukraine is also the first partner country to have participated in the NATO Response Force (NRF), contributing a platoon specialised in nuclear, biological and chemical threats in 2011 and strategic airlift capabilities in 2011. In 2015 and 2016, Ukraine provided strategic airlift, naval and medical capabilities. Currently, Ukraine is contributing with strategic airlift capabilities.

- Lessons learned from hybrid warfare

Against the background of Russia's actions against Ukraine, the NATO-Ukraine Platform on Countering Hybrid Warfare was established at the Warsaw Summit in July 2016. It provides a mechanism to be better able to identify hybrid threats and to build capacity in identifying vulnerabilities and strengthening resilience of the state and society. High-level conferences held under the Platform in Warsaw, Vilnius and Kyiv in 2017 and 2018 have contributed to this objective and raised public awareness. Further projects in support of research, training and expert consultations are ongoing.

- Defence and security sector reform

Ukraine's cooperation with NATO in the area of defence and security sector reform is crucial to the ongoing transformation of Ukraine's security posture and remains an essential part of its democratic transition.

NATO has supported Ukraine's defence and related security sector reform through the Joint Working Group on Defence Reform, and the Planning and Review Process, the NATO Building Integrity Program, the NATO Defence Education Enhancement Program, the Joint Working Group on Defence Technical Cooperation and the advisory mission at the NATO Representation in Kyiv.

Through the Comprehensive Assistance Package for Ukraine endorsed by the NATO-Ukraine Commission at the NATO Summit in Warsaw in 2016, NATO pledged to support Ukraine's goal to implement security and defence sector reforms according to NATO standards by providing strategic-level advice as well as 40 tailored support measures.

A key overarching objective of cooperation in this area is to strengthen democratic and civilian control of Ukraine's armed forces and security institutions, as set out in the Ukrainian Law on National Security of June 2018. Allies contribute to the transformation of Ukraine's defence and security institutions into modern and effective organisations under civilian and democratic control, able to provide a credible deterrence to aggression and defence against military threats. NATO assists Ukraine in the modernisation of its force structure, command and control arrangements, the reform of its logistics system, defence capabilities, and plans and procedures. NATO also provides tailored assistance to strengthen good governance and fight against corruption.

-   The Planning and Review Process

Its participation in the Planning and Review Process (PARP) provides Ukraine with a fundamental mechanism to set realistic reform objectives and to improve its defence and security forces' functional ability to operate alongside Allies in crisis-response operations and other national and international activities to promote security and stability.

The PARP helps guide transformation and reform in the defence and related security sector. The 2018 Partnership Goal package - which sets out goals agreed with the Ukrainian ministries of defence and the interior - explicitly aims to support Ukraine's strategic organisational reforms and institution-building for defence and security sector organisations. This will support Ukraine in pursuing the reforms mandated in its 2015 National Security Strategy and Military Doctrine as well as the Strategic Defence Bulletin approved in 2016. Among the Partnership Goals, 26 are assigned to the defence ministry and the armed forces; 15 to the interior ministry and its subordinate security organisations; and one to cyber defence.

- Capacity-building and civilian control

NATO programs and initiatives contribute to specific aspects of strengthening civilian control over defence and related security institutions, including in the intelligence sector. Improving the capacity of these institutions is of fundamental importance for Ukraine's development as a democratic country. These issues are key deliverables under NATO's Comprehensive Assistance Package for Ukraine and the Partnership Goals agreed under the PARP. As part of wider cooperation in this area, a number of specific initiatives have been taken.

A Professional Development Program (PDP) for civilians working in Ukraine's defence and security institutions was launched in October 2005. The budget for this program was doubled in 2014, with a focus on supporting transformation and reform processes by introducing NATO standards and best practices, building Ukraine's own self-sustained capacity for professional development, and improving inter-agency cooperation and information-sharing. In 2017, the Program also launched implementation of new concepts including the "Champions 100" project providing support to a pool of Ukrainian civil servants directly responsible for Euro-Atlantic integration processes.

In 2007, Ukraine joined the NATO Building Integrity (BI) Program to strengthen integrity, transparency and accountability. The recent completion of the new BI Self-Assessment and Peer Review Process (October 2019) provides a thorough assessment of the previous anti-corruption package and a set of recommendations to improve good governance and pursue sustainable anti-corruption reforms in the defence and related security sectors. On this basis, a tailored program of activities will continue to provide two levels of assistance: specific expertise to the institutions to enhance the good governance and management of defence resources (financial, human and material) and education and training activities to develop individual capacities.

A specific BI educational program to raise awareness on corruption risks and embed BI principles in existing programs of instruction was launched in 2015 with the military and related security institutions of Kharkiv, Khmelnytskyi, Kyiv, Lviv, Odesa and Zhytomyr, as well as the National Defence University of Kyiv. This work is being enhanced through a joint project with the Defence Education Enhancement Program (see below). Additional capacity-building assistance is being provided to civilian institutions like the National Anti-Corruption Bureau, the National Agency for Corruption Prevention as well as representatives of civil society. The BI Program benefits from the expertise and support of the European Union.

- Defence Education Enhancement Program (DEEP)

NATO developed a DEEP program with Ukraine in response to a request from the Ukrainian Defence Minister in 2012. The program is the biggest of its kind with any of NATO's partner countries. It aims to improve and restructure the military education and professional training systems. It focuses specifically on eight defence education institutions in Kharkiv, Kyiv, Lviv, Odesa and Zhytomyr (this includes restoring some Navy Academy capacity in Odesa) and five training centers for Non-Commissioned Officers (NCOs) in Desna, Mykolayv, Starychi, Vasylkyv and Yavoriv.

Additionally, DEEP advises on management of the academies and universities, including supporting faculty on how to teach and development of courses on leadership and decision-making processes. Support has also focused on building e-learning capacity and improving the English language skills of military professors. These efforts expanded to other areas such as organising simulation exercises and courses for demining instructors.

Starting in 2017, DEEP has shifted its assistance into curriculum development in the areas of civilian and democratic control, personnel management, strategic communication, leadership, quality management and NATO operational planning. Following a request from the defence ministry, the program has been extended until 2020.

Training and professionalisation of enlisted soldiers and NCOs is critically important for the success of overall reform in the armed forces. DEEP identified four gap areas in which it now facilitates Allied bilateral support: a) basic soldier combat training; b) training of instructors; c) development of a professional NCO career system; and d) creation of a professional military education for NCOs.

- Military career transition and resettling of former military personnel

NATO supports the reintegration of former military personnel into civilian life through a wide range of projects, adjusted to the new challenges brought up by the Russia-Ukraine conflict. NATO provides concrete assistance in the form of professional retraining and provides psychological rehabilitation services to mitigate post-traumatic stress syndrome among demobilised conscripts. Additionally, NATO is advising on the set-up of an integrated, comprehensive military career transition system through one of the Trust Funds launched at the Wales Summit in 2014 to support security and defence sector reform.

- Destroying stockpiles of weapons and munitions

Individual Allies are supporting the destruction of Ukraine's stockpiles of anti-personnel mines, munitions and small arms and light weapons through Partnership Trust Fund projects. Phase 1 of the Trust Fund led by the United States involved the safe destruction of 400,000 small arms and light weapons (SALW), 15,000 tons of munitions and 1,000 man-portable air defence systems (MANPADS) in the 2006-2011 timeframe. A second phase started in 2012. As of May 2018, it has successfully destroyed more than 130,000 SALW, 27,200 tonnes of conventional ammunition and 1.7 million anti-personnel landmines. Its scope was extended in 2017-2018 to support enhanced ammunition safety management.

Another Trust Fund led by Germany supports the disposal of radioactive waste from former Soviet military sites in Ukraine. A project enabling Ukraine to recover and secure radioactive material according to international standards and to restore the site to its original condition was carried out in 2016-17. A follow-on project was launched in December 2017.

- Air Situation Data Exchange (ASDE)

Ukraine joined the ASDE program in July 2006. Through the exchange of filtered air situation information it reduces the risk of potential cross-border incidents and optimises responses to terrorist attacks using civil airplanes. Connections between NATO and Ukraine have been in operation via Hungary since end 2008 and via Turkey since mid-2011. Following the Russia-Ukraine crisis, air data information provided by NATO has been extended to cover a larger area. Work is ongoing to provide Ukraine with a connection for the transfer of classified information.

- Trust Funds promoting security and defence sector reform and capability development

At the Wales Summit in 2014, Allies decided to launch substantial new programs to enhance NATO's assistance to capability development and sustainable capacity-building in Ukraine's security and defence sector. Six Trust Funds were set up, making use of a mechanism which allows individual Allies and partner countries to provide financial support for concrete projects on a voluntary basis. Subsequently, all Allies have contributed in one way or the other to the development of these Trust Funds. They include:

- Trust Fund on Command, Control, Communications and Computers (C4)

The C4 Trust Fund assists Ukraine in reorganising and modernising its C4 structures and capabilities, facilitates their interoperability with NATO to contribute to NATO-led exercises and operations, and enhances Ukraine's ability to provide for its own defence and security.

The Trust Fund is led by Canada, Germany and the United Kingdom, with the NATO Communications and Information Agency as executing agent. NATO conducted a C4 Feasibility Study to assess Ukraine's capabilities and needs based on fact-finding trips to Ukraine to identify priority C4 requirements through consultations with Ukrainian authorities. A final report on recommendations for reform, reorganisation and modernisation of Ukraine's Armed Forces and capabilities in the C4 area was delivered in September 2016.

Four projects have been agreed and are at various stages of implementation:

o Regional Airspace Security Program (RASP): to promote regional airspace security cooperation and interoperability with NATO, improve Ukraine's internal civil-military airspace cooperation, and to establish cross-border coordination capability with Allies for better handling of air security incidents. The project will be implemented by Spring 2020 and a follow-up project to allow for continued support is under development.

o Secure Tactical Communications Project: to assist Allies in providing secure communications equipment to enhance Ukraine's capabilities for secure command and control and situational awareness for its armed forces. The project has been implemented, providing a resilient communications capability at the tactical level. A follow-up project to provide additional equipment is under development.

o Knowledge Sharing: to provide NATO subject-matter expertise, documentation, training, standards, best practices, mentoring and advice to C4 project teams and subject-matter experts in Ukraine. The first phase of this project has been finalised and a follow-up project is currently being implemented.

o Situational Awareness (SA): to provide a reference for the implementation of NATO standards, software SA tools (JOCWatch, JCHAT, iGeoSit), procedures as well as support through mentoring and subject-matter experts for Ukrainian capability development.

- Trust Fund on Logistics and Standardization

This Trust Fund aims to support the ongoing reform of Ukraine's logistics and standardisation systems for the armed forces as well as other national military formations, including the National Guard and the State Border Security Service, as appropriate.

Led by the Czech Republic, the Netherlands and Poland, the project builds on the findings of a Strategy Level Gap Analysis conducted in the course of 2015.

It complements and is aligned with other NATO activities performed in these areas such as those under the Planning and Review Process, Joint Working Group on Defence Technical Cooperation and Joint Working Group on Defence Reform.

Over the course of three years, the project aims will be achieved through the implementation of three capability-driven initiatives in support of long-term developments, with a focus on National Codification Capability Enhancement, Supply Chain Management Capability Improvement, and Standardization Management Capability Improvement.

- Trust Fund on Cyber Defence

This Trust Fund, led by Romania, aimed to help Ukraine develop strictly defensive, technical capabilities to counter cyber threats. Assistance included the establishment of an incident management center for monitoring cyber security incidents and laboratories to investigate cyber security incidents. The project was completed in 2017.

Current efforts focus on identifying follow-up activities in the area of cyber defence.

-   Trust Fund on Medical Rehabilitation

This Trust Fund aims to ensure that patients, active and discharged Ukrainian servicemen and women and civilian personnel from the defence and security sector, have rapid access to appropriate care. Furthermore, it seeks to support Ukraine in enhancing its medical rehabilitation system to ensure that long-term sustainable services are provided.

The project, led by Bulgaria and executed by the NATO Support and Procurement Agency (NSPA), started in 2015 and runs over 48 months.

As of May 2018, the medical rehabilitation of 270 servicemen has been supported and 15 servicemen have been provided with prostheses. Support to an additional 100 servicemen will be provided in 2018-2019. Moreover, 13 servicemen from the defence ministry have benefited from vocational rehabilitation services. Another 148 former servicemen and 140 civilians/internally displaced persons from the Donbas have accessed rehabilitation through sport. In partnership with the NATO-sponsored project on resettling former military personnel, more than 6,000 former servicemen have benefited from psychological support services. In 2017, support was also provided for Ukraine to participate at their first Invictus Games.

Five medical rehabilitation units in hospitals have received appropriate equipment to improve the quality of services. The first occupational therapy kitchen in Ukraine, the first modern rehabilitation swimming pool and the first wheelchair workshop in a governmental institution were delivered in 2016. More than 2,200 Ukrainian physical and psychological professionals from the medical rehabilitation sector, both from government and non-governmental organisations , have benefitted from professional development activities. Since March 2018, the Trust Fund is also supporting the development of internationally recognised academic curricula for prosthetists/orthotists and orthopedic technologists, professions newly recognised in 2016.

- Trust Fund on Military Career Transition

This Trust Fund, led by Norway, assists Ukraine in developing and implementing a sustainable, effective and integrated approach to the resettlement of military personnel embedded in the personnel management of the armed forces.

The project aims to increase understanding among Ukrainian officials of the main organisation and managerial concepts of social adaptation systems, and develop their professional skills. It will also help define parameters for the assistance for resettlement within the armed forces through a combination of seminars, workshops, study tours and analytical surveys.

Trust Fund on Explosive Ordnance Disposal (EOD) and Countering Improvised Explosive Devices (C-IED)

This Trust Fund, led by Slovakia, directly supports specific selected recommendations in the NATO EOD and C-IED Assistance Plan to Ukraine. The project selectively supports civil humanitarian activities in the clearance of explosive hazards, including IEDs. It will assist in setting the foundations for transformation of EOD and development of C-IED in Ukraine based on NATO policy and practice, particularly regarding multi-agency cooperation. Three primary initiatives covering doctrine, interoperability and civil support will be initiated over a two-year period, starting in summer 2018.

- Military-to-military cooperation today

Helping Ukraine implement its defence reform objectives is also a key focus of military-to-military cooperation, complementing the work carried out under the Joint Working Group on Defence Reform and the Planning and Review Process with military expertise.

Another important objective is to develop operational capabilities and interoperability with NATO forces through a wide range of activities and military exercises. These exercises allow military personnel to gain hands-on experience in working with forces from NATO countries and other partners. Ukraine is part of the Partnership Interoperability Initiative, launched at the 2014 Wales Summit, which aims to maintain the levels of interoperability developed by international forces serving in the NATO-led International Security Assistance Force in Afghanistan (2003-2014).

An important part of practical military-to-military cooperation is carried out under the Military Committee with Ukraine Work Plan, making use of the educational, training, exercise, assistance, and advisory activities which NATO offers to partner countries. The Military Partnerships Directorate of Allied Command Operations is responsible for the Work Plan's implementation. All these activities focus on improving the interoperability and reinforcing the operational capabilities of Ukraine's armed forces, but also substantially contribute to ongoing security and defence reform.

Ukraine's active participation in the NATO Operational Capabilities Concept Evaluation and Feedback Program supports the further development of the armed forces, while also enabling the Alliance to put together tailored force packages that can be deployed in support of NATO-led operations and missions.

The military side has also taken the lead in developing a legal framework to enable NATO and Ukraine to further develop operational cooperation:

- o A Partnership for Peace (PfP) Status of Forces Agreement facilitates participation in PfP military exercises by exempting participants from passport and visa regulations and immigration inspection on entering or leaving the territory of the country hosting the event (entered into force in May 2000).

- o A Host Nation Support Agreement addresses issues related to the provision of civil and military assistance to Allied forces located on, or in transit through, Ukrainian territory in peacetime, crisis or war (ratified in March 2004).

- o A Strategic Airlift Agreement enables Ukraine to make a substantial contribution to NATO's capability to move outsized cargo by leasing Antonov aircraft to Allied armed forces, an arrangement which also brings economic benefits to Ukraine (ratified in October 2006).

Senior Ukrainian officers also regularly participate in courses at the NATO Defence College in Rome, Italy and the NATO School in Oberammergau, Germany. Contacts with these establishments have been instrumental in setting up a new multinational faculty at the Ukrainian Defence Academy.

- Defence technical cooperation

Defence technical cooperation focuses on enhancing the interoperability of Ukrainian contributions to international operations with the forces of NATO nations.

Cooperation in this area began with the entry of Ukraine to the Partnership for Peace and, in particular, their participation in a number of groups that meet under the auspices of the Conference of National Armaments Directors (CNAD), the senior NATO body responsible for promoting cooperation between Allies and partners in the armaments field.

Since December 2014, Ukraine's participation in the CNAD and its substructure has been taking place in the context of the Interoperability Platform. The CNAD identifies opportunities for cooperation between nations in capability development, defence equipment procurement processes and the development of technical standards.

The Joint Working Group on Defence Technical Cooperation, which met for the first time in March 2004, works toward increased cooperation in this area between

NATO and Ukraine on the basis of a Roadmap on Defence Technical Cooperation. Current priorities include:

- o Standardisation and codification as a means for increasing interoperability of the Ukrainian armed forces with Allied forces.

- o Implementation of the Trust Fund projects on command, control, communications and computers (C4) and demilitarisation of expired ammunition and excess small arms and light weapons.

- o Cooperation in the framework of the CNAD and with the NATO Science and Technology Organization.

- o Sharing of expertise in the area of life cycle management, with Ukraine having hosted a related workshop in July 2019.

- o Ukraine's participation in NATO's Smart Defence projects, including "Malware Information Sharing Platform", "NATO Multinational Cyber defence education and training", "Flexible and interoperable Toolbox meeting the future operational requirements in confined and shallow waters", and "Female Leaders in Security and Defence".

- o Support provided to Ukraine in the modernisation of their defence industry.

- o Continued use and enhancement of the Air Situation Data Exchange (ASDE) program.

- Civil preparedness

Civil preparedness remains an important driver of NATO cooperation with Ukraine. Since the start of the 2014 crisis in Crimea and eastern Ukraine, it has been at the forefront of Alliance solidarity and support. In April 2014, a team of civil experts visited Kyiv to provide advice on Ukraine's contingency plans and crisis-management measures related to critical energy infrastructure and civil protection risks.

Today, NATO-Ukraine cooperation in the area of civil preparedness focuses on improving national capacity for civil preparedness and resilience in facing hybrid threats through the exchange of lessons learned, best practices and the provision of expert advice.

Ukraine is also a regular participant in disaster preparedness and response exercises organised by the Euro-Atlantic Disaster Response Coordination Center (EADRCC). Ukraine already hosted three such exercises in 2000, 2005 and 2015. The 2015 EADRCC exercise, which was inaugurated by NATO Secretary General Jens Stoltenberg and President Petro Poroshenko, was one of the largest field exercises organised by the EADRCC, with over 1,100 participants from 26 Allies and partner countries.

NATO's Science for Peace and Security (SPS) Program:

Active engagement between Ukraine and the SPS Program dates back to 1991. A Joint Working Group on Scientific and Environmental Cooperation oversees cooperation in this area. In April 2014, in response to the crisis in Ukraine, practical cooperation with Ukraine in the field of security-related civil science and technology has been further enhanced.

Today, SPS activities in Ukraine address a wide variety of emerging security challenges such as counter-terrorism, advanced technologies, cyber defence, energy security, and defence against chemical, biological, radiological and nuclear (CBRN) agents. SPS activities also deal with human and social aspects of security, such as the implementation of United Nations Security Council Resolution 1325 on Women, Peace and Security; support the development of advanced technologies with security applications; and assist with the detection and clearance of mines and unexploded ordnance. Many current activities help Ukraine to deal with the negative effects of the crisis, engaging Allied and Ukrainian scientists and experts in meaningful, practical cooperation, forging networks and supporting capacity-building in the country.

Through tailored capability and capacity-building measures, the SPS Program provided support to the Comprehensive Assistance Package (CAP) for Ukraine, endorsed at the 2016 NATO Summit in Warsaw. One important project assisted Ukraine in the area of humanitarian demining by enhancing the capacity of the State Emergency Service of Ukraine (SESU) in undertaking demining operations in eastern parts of the country. Through this project, the SPS Program was also able to immediately respond to an urgent request for equipment following the Balaklia Arms Depot explosion in Ukraine in March 2017. Furthermore, an ongoing multi-year initiative for the development of a 3D mine detector will ensure the sustainability of the activities.

The Program also helped to build capacity in the sphere of telemedicine and paramedicine in the framework of the CAP. As part of the project, two paramedic centers in Ukraine were equipped and 30 Ukrainian paramedics took part in a "train-the-trainer" course in Romania. During the Euro-Atlantic Disaster Response Coordination Center's (EADRCC) field exercise in Tuzla, Bosnia and Herzegovina, in 2017, the telemedicine capabilities were successfully live-tested, allowing medical specialists to engage in disasters or incidents across national borders. The project has now been brought to a successful completion.

Ukrainian experts are also involved in a significant new SPS initiative developing innovative technologies for the stand-off detection of explosives, a project that is contributing to NATO's enhanced role in the international fight against terrorism. Ukraine is a key contributor to the SPS flagship program DEXTER (Detection of EXplosives and firearms to counter TERrorism) along with eight Allies and partner countries developing an integrated system that can detect explosives and firearms in public places, remotely and in real time without disrupting the flow of pedestrians.

Since 2014, Ukraine has been the largest beneficiary of the SPS Program in line with the Allies 'decision to intensify cooperation with Ukraine. As a result, in the past five years, a total of 69 SPS activities with Ukraine as leading partner were launched, with Ukrainian scientists and experts participating in a number of additional SPS-supported projects or workshops as researchers or speakers. In 2019, Ukraine remains the lead partner with 32 ongoing SPS projects, which represents 17 per cent of the overall SPS Program.

The importance of the SPS cooperation program with Ukraine was highlighted, in November 2018, at NATO Headquarters, on the occasion of the 60th anniversary of the NATO Science Program. A Ukrainian professor of the National Academy of Science received the SPS Partnership Prize for the Compact Sensor for Unmanned Aerial Vehicles Project.

# NATO-EUROPEAN UNION COOPERATION

The North Atlantic Treaty Organization and the European Union share strategic interests and the EU is a unique and essential partner for NATO. The two organisations also share a majority of members, and all members of both organisations share common values.

So, both NATO and the European Union have, since their inception, contributed to maintaining and strengthening security and stability in Western Europe. NATO has pursued this aim in its capacity as a strong and defensive political and military alliance and, since the end of the Cold War, has extended security in the wider Euro-Atlantic area both by enlarging its membership and by developing other partnerships.

The European Union has created enhanced stability by promoting progressive economic and political integration, initially among western European countries and subsequently also by welcoming new member countries. As a result of the respective organisations' enlargement processes, an increasing number of European countries have become part of the mainstream of European political and economic development, and many are members of both organisations.

This way, no formal relationship existed between NATO and the European Union until 2000. Prior to that, during the 1990s the Western European Union (WEU) acted as the interface for cooperation between NATO and those European countries seeking to build a stronger European security and defence identity within NATO.

Recall that at the turn of the 21st century, after the end of the Cold War, WEU tasks and institutions were gradually transferred to the European Union (EU), providing central parts of the EU's new military component, the European Security and Defence Policy (ESDP). This process was completed in 2009 when a solidarity clause between the member states of the European Union, which was similar (but not identical) to the WEU's mutual defence clause, entered into force with the Treaty of Lisbon. The states party to the Modified Treaty of Brussels consequently decided to terminate that treaty on 31 March 2010. On 30 June 2011, the WEU was officially declared defuncted.

But the situation changed fundamentally in 1999 when, against the backdrop of the conflicts in the Balkans, EU leaders decided to develop a European Security and Defence Policy within the European Union itself, in coordination with NATO, and to take over responsibility for most of the functions that had been exercised by the Western European Union. The following year, NATO and the European Union started to work together to develop a framework for cooperation and consultation. This led to the development of a strategic partnership (NATO-EU Declaration on the European Security and Defence Policy (ESDP) between the two organisations and the agreement of the Berlin Plus arrangements, which provide access to NATO's collective assets and capabilities for military operations led by the European Union.

Thus, these developments established the basis for NATO-EU cooperation in the sphere of crisis management in the western Balkans as well as for the development of cooperation on other issues.

**The evolution of NATO-EU relations**

The Cold War period:

Let us start by saying that despite shared objectives and common interests in many spheres, the parallel development of NATO and the European Union throughout the Cold War period was characterised by a clear separation of roles and responsibilities, and the absence of formal or informal institutional contacts between them. While a structural basis for a specifically European security and defence role existed in the form of the Western European Union, created in 1948, for practical purposes western European security was preserved exclusively by NATO.

For its part, the Western European Union undertook a number of specific tasks, primarily in relation to post-war arms control arrangements in Western Europe. However, its role was limited and its membership was not identical to that of the European Union.

Just like that, given this institutional background, when questions arose concerning the need for a more equitable sharing of the burden of European security between the two sides of the Atlantic, they were discussed primarily at the bilateral, political level. A number of representational initiatives on the part of the European member countries of NATO were conducted with a view to reassuring the United States about the level of the European commitment to security and defence. However, no multilateral or institutional arrangements existed for developing independent structures outside the Alliance framework.

NATO-WEU cooperation and the development of a European Security and Defence Identity within NATO:

It became apparent that European countries needed to assume greater responsibility for their common security and defence in the early 1990s. A rebalancing of the relationship between Europe and North America was essential for two reasons: first, to redistribute the economic burden of providing for Europe's continuing security, and second, to reflect the gradual emergence within European institutions of a stronger, more integrated European political identity, and the conviction of many EU members that Europe must develop the capacity to act militarily in appropriate circumstances where NATO is not engaged militarily.

The emergence of these new approaches to the problems of European security was profoundly influenced by the conflicts in the western Balkans during the 1990s. The inability of Europe to intervene to prevent or resolve such conflicts led to a collective realisation that the European Union must redress the imbalance between its far-reaching economic power and the limitations on its political power. It had become obvious to many that a coordinated diplomatic effort to end conflict by political means needed to be backed up, if necessary, by credible military force. This led the European Union to become increasingly committed during the 1990s to conflict prevention and crisis management beyond its borders.

Thereby, an important step in this direction was taken in 1992 with the Treaty of Maastricht, which included an agreement by EU leaders to develop a Common Foreign and Security Policy (CFSP) "including the eventual framing of a common defence policy which might in time lead to a common defence". As an integral part of the development of the European Union, the Western European Union was requested to elaborate and implement EU decisions and actions with defence implications. The initiative to develop European defence capabilities through the Western European Union was later carried forward on the basis of the 1997 EU Treaty of Amsterdam. This Treaty, which entered into force in May 1999, incorporated the so-called WEU Petersburg tasks (humanitarian search and rescue missions, peacekeeping missions, crisis management tasks including peace enforcement, and environmental protection) providing the basis for the operative development of a common European defence policy.

And in the same timeframe, a decision was taken at the 1994 NATO summit meeting in Brussels to develop a European Security and Defence Identity within NATO. This led to the introduction of practical arrangements to enable the Alliance to support European military operations undertaken by the Western European Union. Decisions which served to reinforce this development were taken by the Alliance at subsequent meetings of NATO foreign and defence ministers in Berlin and Brussels in June 1996, and at the 1997 NATO summit meeting in Madrid.

So, the Western European Union was simultaneously developed as the defence component of the European Union and as a means of strengthening the European pillar of NATO. European member countries of the Alliance recognised that in the process of achieving a genuine European military capability, unnecessary duplication of the command structures, planning staffs and military assets and capabilities already available within NATO should be avoided. Moreover, such an approach would serve to strengthen the European contribution to the Alliance's missions and activities, while responding to the European Union's goal of developing a common foreign and security policy as well as to the overall need for a more balanced transatlantic partnership.

In this sense, the arrangements made for NATO-WEU cooperation from 1991 to 2000 laid the groundwork for the subsequent development of the future NATO-EU relationship. In practice these arrangements were designed to ensure that if a crisis arose in which the Alliance decided not to intervene but the Western European Union chose to do so, the WEU could request the use of Alliance assets and capabilities to conduct an operation under its own political control and strategic direction.

Thus, in December 1998, new impetus for the development of this relationship was provided by the British-French summit meeting at St Malo. France and the United Kingdom agreed that the European Union "must have the capacity for autonomous action, backed up by credible military forces, the means to decide to use them, and a readiness to do so, in order to respond to international crises". They issued a joint statement outlining their determination to enable the European Union to give concrete expression to these objectives.

Transfer of WEU responsibilities to the European Union:

It is important to remember that in the new climate that prevailed after the Anglo-French initiative, further decisions could be made. At the Washington Summit in April 1999, NATO leaders welcomed the new impetus given to the strengthening of the European Security and Defence Policy, affirming that a stronger European role would help contribute to the vitality of the Alliance in the 21st century. NATO leaders further stated that as this process went forward, NATO and the European Union should ensure the development of effective mutual consultation, cooperation and transparency, building on existing mechanisms between NATO and the Western European Union.

In the same way we want to review that they also set in train further work to address a number of principles for future cooperation with the European Union and, in particular, satisfactory resolution of outstanding questions. These related in particular to three issues that had long proved difficult to resolve, namely:

   a. the means of ensuring the development of effective mutual consultation, cooperation and transparency between the European Union and the Alliance, based on the mechanisms that had been established between NATO and the Western European Union;

   b. the participation of non-EU European Allies in the decisions and the operations that might be conducted by the European Union; and

   c. practical arrangements for ensuring EU access to NATO planning capabilities and NATO's assets and capabilities.

EU leaders meeting in Cologne in June 1999 welcomed the St Malo statement and, taking into account the Amsterdam Treaty which incorporated the WEU Petersburg tasks, agreed on the concept and the objective of a European

Security and Defence Policy (ESDP) for the European Union.

They decided "to give the EU the means and capabilities to assume its responsibilities regarding a common European policy on security and defence" and also made a commitment to ensure the development of effective mutual consultation, cooperation and transparency with NATO. Similar reassurances were offered at subsequent European Council meetings, particularly in Helsinki (December 1999) and Nice (December 2000).

Otherwise, at the Helsinki meeting, the European Council established a "Headline Goal" for EU member states in terms of developing military capabilities for crisis-management operations. Its objective was to enable the European Union, by 2003, to deploy and sustain for at least one year military forces of up to 60 000 troops to undertake the full range of the Petersburg tasks referred to above, in the context of EU-led military operations in response to international crises where NATO as a whole is not engaged militarily.

In addition, the European Union decided to create permanent political and military structures, including a Political and Security Committee, a Military Committee and a Military Staff, to ensure the necessary political guidance and strategic direction for such operations.

The crisis management role of the Western European Union was also transferred to the EU at the Helsinki meeting (decision taken at the WEU Council Ministerial meeting in Marseilles in December 2000). The residual responsibilities of the WEU remain unaffected and are handled by a much reduced formal political structure and a small secretariat.

In addition to this, built on decisions taken in Cologne and Helsinki, the Treaty of Nice signed in December 2000 (which came into effect in February 2003) provides the EU with the political framework for military operations (ESDP) and permanent political and military structures.

So, at the end of 2000, with the formal transfer of responsibilities for EU decisions and actions with defence implications from the Western European Union to the European Union itself, the relationship between NATO and the EU took on a new dimension.

Towards a strategic partnership with the European Union:

In this truck, negotiations initiated in September 2000 led to an exchange of letters between NATO's Secretary General and the EU Presidency in January 2001 to define the scope of cooperation and the modalities of "consultations and cooperation on questions of common interest relating to security, defence and crisis management, so that crises can be met with the most appropriate military response and effective crisis management ensured".

Thus, the exchange of letters provided for joint meetings at different levels. It prescribed two joint NATO-EU foreign ministers meetings every year and a minimum of three joint meetings per semester at ambassadorial level of the North Atlantic Council and the EU Political and Security Committee (known as NAC-PSC meetings). In addition, two joint Military Committee meetings would be held each semester, and meetings between subordinate committees would be scheduled on a regular basis. The exchange of letters also provided for meetings at staff level.

And since then, NAC-PSC meetings have become a normal feature of cooperation between the two organisations. The September 2001 terrorist attacks in the United States provided a further incentive to enhance cooperation. The very next day, NATO's Secretary General participated in the deliberations of the EU General Affairs Council to analyse the international situation following the attacks. Formal contacts and reciprocal participation in meetings have subsequently increased.

In November 2002, at the Prague Summit, NATO leaders reaffirmed their commitment to enhance NATO-EU cooperation, the effectiveness of which had already been evident in joint efforts to restore peace and create the conditions for progress in the Balkans.

In December 2002, the NATO-EU Declaration on ESDP issued, the two organisations "welcomed the strategic partnership established between the EU and NATO in crisis management, founded on our shared values, the indivisibility of our security and our determination to tackle the challenges of the new century".

A few months later, NATO and the European Union gave substance to this strategic partnership and opened the way for coordinated action by agreeing a series of documents that provided for exchanges of classified information and for cooperation in crisis management, including through the Berlin Plus arrangements.

## The development of practical NATO-EU cooperation

The Berlin Plus arrangements:

We want to emphasise that the Berlin Plus arrangements are based on the recognition that member countries of both organisations only have one set of forces and limited defence resources on which they can draw. Under these circumstances, and to avoid an unnecessary duplication of resources, it was agreed that operations led by the European Union would be able to benefit from NATO assets and capabilities. In effect, these arrangements enable NATO to support EU-led operations in which the Alliance as a whole is not engaged. They have facilitated the transfer of responsibility from NATO to the European Union of military operations in the former Yugoslav Republic of Macedonia and in Bosnia and Herzegovina.

These arrangements, agreed in March 2003, are referred to as Berlin Plus because they build on decisions taken in Berlin in 1996 in the context of NATOWEU cooperation. The main features of the Berlin Plus arrangements consist of the following main elements:

    a.  assured EU access to NATO planning capabilities able to contribute to military planning for EU-led operations;

    b.  the presumption of availability to the European Union of pre-identified NATO capabilities and common assets for use in EU-led operations;

    c.  identification of a range of European command options for EU-led operations, further developing the role of NATO's Deputy Supreme Allied Commander Europe (DSACEUR) in order for him to assume his European responsibilities fully and effectively;

d.  the further adaptation of NATO's defence planning system to incorporate more comprehensively the availability of forces for EU-led operations;

e.  a NATO-EU agreement covering the exchange of classified information under reciprocal security protection rules;

f.  procedures for the release, monitoring, return and recall of NATO assets and capabilities;

g.  NATO-EU consultation arrangements in the context of an EU-led crisis management operation making use of NATO assets and capabilities.

Cooperation in the western Balkans:

The emergency in Serbia and the unstable political situation in the former Yugoslav Republic of Macedonia became a focus of international concern in 2001. A series of joint visits to the region by NATO's Secretary General and the EU High Representative for Common Foreign and Security Policy underscored the unity of purpose and commitment shared by NATO and the European Union with regard to the security of the region.

At the first formal NATO-EU foreign ministers' meeting in Budapest on 30 May 2001, the NATO Secretary General and the EU presidency issued a joint statement on the western Balkans. Later, they met in Brussels in December

2001 and in Reykjavik in May 2002 to review their cooperation across the board. They underlined their continuing commitment to strengthening the peace process in the former Yugoslav Republic of Macedonia as well as elsewhere in the western Balkans, and reaffirmed their commitment to a close and transparent relationship between the two organisations.

On the other hand, cooperation on the ground contributed positively to the improved situation in the former Yugoslav Republic of Macedonia. From August 2001 to the end of March 2003, NATO provided security for EU and OSCE monitors of the peace plan brokered with the support of the international community in the city of Ohrid. On 31 March 2003, the NATO-led peacekeeping mission (Operation Amber Fox) was terminated and responsibility for this task was formally handed over to the European Union, with the agreement of the government in Skopje. Renamed Operation Concordia, this was the first EU-led military crisis-management operation. Undertaken on the basis of the Berlin Plus arrangements, it marked the real starting point for cooperation between NATO and the European Union in addressing an operational crisis-management task.

NATO and the European Union, on 29 July 2003, formally agreed on a "concerted approach to security and stability in the western Balkans" and outlined their strategic approach to the problems of the region. Both organisations expressed determination to continue to build on their achievements in working together to bring an end to conflict and to help stabilise the region as a whole.

In June 2004, at the Istanbul Summit, in view of the positive evolution of the security situation in Bosnia and

Herzegovina, Alliance leaders confirmed their decision to terminate NATO's peacekeeping mission there, which it had led since 1996, and welcomed the readiness of the European Union to assume responsibility for a new mission, Operation Althea, based on the Berlin Plus arrangements. Close cooperation and coordination with regard to the planning and implementation of the EU mission was facilitated by the appointment of the NATO Deputy Supreme Allied Commander Europe (DSACEUR) as the EU Operation Commander.

Thereby, NATO leaders stressed that NATO would nevertheless remain committed to the stabilisation of the country and would maintain a residual military presence through a NATO headquarters in Sarajevo. This headquarters is responsible primarily for providing assistance in the defence reform process and other tasks including counter-terrorism and support for the International Criminal Tribunal for the former Yugoslavia.

On 2 December 2004, a ceremony marking the handover of the primary responsibility for security in Bosnia and Herzegovina from NATO to the European Union took place in Sarajevo. The new NATO military headquarters was formally established on the same day.

Cooperation on other issues:

Just like that, the strategic partnership also covers other issues of common interest. These include concerted efforts with regard to the planning and development of military capabilities. NATO experts have provided military and technical advice for both the initial preparation and the subsequent implementation of the European Union's

European Capabilities Action Plan (ECAP), which was created in November 2001. ECAP aims to provide the forces and capabilities required to meet the EU Headline Goal set at Helsinki in 1999. A NATO-EU Capability Group, established in May 2003, is working to ensure that the Alliance's capabilities initiatives and the ECAP are mutually reinforcing and is also examining the relationship between the NATO Response Force and newly created EU Battle Groups, as part of the NATO-EU agenda under the Berlin Plus arrangements.

Moreover, the EU rapid reaction units, composed of battle groups, were part of the new Headline Goal for 2010 announced in February 2004. They are to be completely developed by 2007. The Headline Goal also led to the creation of an EU Defence Agency that focuses on the development of defence capabilities, research, acquisition and armaments.

Equally, through information exchanges on their respective activities, consultations and contacts at expert and staff level, and joint meetings, NATO and the European Union also undertake joint work on issues such as the fight against terrorism, the proliferation of weapons of mass destruction (WMD), the situation in Moldova, Mediterranean issues and cooperation in Afghanistan. Additional spheres of information exchange and cooperation include protection of civilian populations against chemical, biological, radiological and nuclear attacks and other civil emergency planning and WMD-related issues. Cooperation can sometimes involve reciprocal participation in exercises. In November 2003 for instance, the first joint NATO-EU crisis management exercise (CME/CMX 03) was held. It was based on a range of standing Berlin Plus arrangements and concentrated on how the EU plans for an envisaged EU-led operation with recourse to NATO assets and capabilities, where NATO as a whole is not engaged.

**Fundamental aspects of NATO-EU cooperation today**

Sharing strategic interests and facing the same challenges, NATO and the European Union (EU) cooperate on issues of common interest and are working side by side in crisis management, capability development and political consultations, as well as providing support to their common partners in the east and south. The EU is a unique and essential partner for NATO. The two organisations share a majority of members, have common values and face similar threats and challenges.

Strengthening the NATO-EU strategic partnership is particularly important in the current security environment, in which both organisations and their members are facing the same challenges to the east and south.

In Warsaw in July 2016, Allied leaders underlined that the EU remains a unique and essential partner for NATO. Enhanced consultations at all levels and practical cooperation in operations and capability development have brought concrete results. The security challenges in the two organisations 'shared eastern and southern neighbourhoods make it more important than ever to reinforce the strategic partnership.

Allied leaders welcomed the joint declaration issued in Warsaw by the NATO Secretary General, the President of the European Council and the President of the European

Commission, which outlines a series of actions the two organisations intend to take together in concrete areas, including countering hybrid threats , enhancing resilience, defence capacity building, cyber defence, maritime security, and exercises.

As a follow-up, in December 2016, NATO foreign ministers endorsed 42 measures to advance how NATO and the EU work together, including:

- measures to bolster resilience to hybrid threats, ranging from disinformation campaigns to acute crises;

- cooperation between NATO's Operation Sea Guardian and the EUNAVFOR Operation Sophia in the Mediterranean;

- exchange of information on cyber threats and the sharing of best practices on cyber security;

- ensuring the coherence and complementarity of each other's defence planning processes;

- parallel and coordinated exercises;

- efforts to support the local capacities of partner countries in the sectors of security and defence.

In December 2017, further steps were taken to boost NATO-EU cooperation through the addition of 32 new measures including in three new areas:

- military mobility to ensure that forces and equipment can move quickly across Europe if needed, which requires procedures for rapid border crossing, sufficient transport assets and robust infrastructure (roads, railways, ports and airports);

- information-sharing in the fight against terrorism and strengthening coordination of counter-terrorism support for partner countries;

- promoting women's role in peace and security.

On 10 July 2018, just ahead of the NATO Summit in Brussels, the two organisations underlined the essential nature of continued cooperation to address multiple and evolving security challenges. They agree to focus on swift progress in the areas of military mobility, counter-terrorism and strengthening resilience to chemical, biological, radiological and nuclear-related risks as well as promoting the women, peace and security agenda. They also welcome the complementary and mutually reinforcing efforts of the EU and NATO to strengthen capabilities in defence and security.

At the July 2018 Brussels Summit, Allied leaders welcomed the joint declaration and the tangible results achieved in a range of areas, underlining their determination to develop and deepen cooperation by fully implementing the common set of 74 proposals. They recognised that the development of coherent, complementary and interoperable defence capabilities, avoiding duplication, is key to joint efforts to make the Euro-Atlantic area safer. Such efforts will lead to a stronger NATO, help enhance common security, contribute to transatlantic burden sharing, help deliver needed capabilities, and support an overall increase in defence spending.

Allied leaders also welcomed the call for deeper NATO-EU cooperation by the European Council in June 2018. They noted that the fullest participation of non-EU Allies in the EU's efforts to strengthen their capacities to address common security challenges is essential for the strengthened strategic partnership between NATO and the EU.

The NATO Secretary General and the EU High Representative regularly report to NATO Allies and EU member states on progress in cooperation. (See reports: June 2017, November 2017, June 2018, June 2019)

Non-EU European Allies make a significant contribution to these efforts. For the strategic partnership between NATO and the EU, their fullest involvement in these efforts is essential.

NATO and the EU can and should play complementary and mutually reinforcing roles in supporting international peace and security. The Allies are determined to make their contribution to create more favourable circumstances through which they will:

- fully strengthen the strategic partnership with the EU, in the spirit of full mutual openness, transparency, complementarity and respect for the autonomy and institutional integrity of both organisations;

- enhance practical cooperation in operations throughout the crisis spectrum, from coordinated planning to mutual support in the field;

- broaden political consultations to include all issues of common concern, in order to share assessments and perspectives;

- cooperate more fully in capability development, to minimise duplication and maximise cost effectiveness.

Cooperation in the field:

A) Combating illegal trafficking in humans in the Aegean and the Central Mediterranean

- NATO defence ministers decided on 11 February 2016 to deploy ships to the Aegean Sea to support Greece and Turkey, as well as the European Union's border agency Frontex in their efforts to tackle the refugee and migrant crisis. Standing NATO Maritime Group 2 (SNMG2) is conducting reconnaissance, monitoring and surveillance in the territorial waters of Greece and Turkey, as well as in international waters. The deployment in the Aegean Sea aims to support international efforts to cut the lines of human trafficking and illegal migration. NATO ships are providing real-time information to the coastguards and relevant national authorities of Greece and Turkey, as well as to Frontex, helping them in their efforts to tackle this crisis.

- In October 2016, ministers agreed to extend NATO's deployment in the Aegean Sea and also decided that NATO's new Operation Sea Guardian will support the EU's Operation Sophia in the Central Mediterranean with NATO ships and planes, ready to help increase the EU's situational awareness and provide logistical support.

B)  The Western Balkans

- In July 2003, the EU and NATO published a " Concerted Approach for the Western Balkans ". Jointly drafted, it outlines core areas of cooperation and emphasises the common vision and determination both organisations share to bring stability to the region.

- The Republic of North Macedonia

On 31 March 2003, the EU-led Operation Concordia took over the responsibilities of the NATO-led mission, Operation Allied Harmony, in the country at the time known as the former Yugoslav Republic of Macedonia. This mission, which ended in December 2003, was the first "Berlin Plus" operation in which NATO assets were made available to the EU.

- Bosnia and Herzegovina

Building on the results of Concordia and following the conclusion of the NATO-led Stabilisation Force (SFOR) in Bosnia and Herzegovina, the EU deployed a new mission called Operation Althea on

2 December 2004. The EU Force (EUFOR) operates under the "Berlin Plus "arrangements, drawing on NATO planning expertise and on other Alliance's assets and capabilities. The Vice-Chief of Staff SHAPE is the Commander of Operation Althea. The EU Operation Headquarters (OHQ) is located at SHAPE.

- Kosovo

NATO has been leading a peacekeeping force in Kosovo (KFOR) since 1999. The EU has contributed civil assets to the UN Mission in Kosovo (UNMIK) for years and agreed to take over the police component of the UN Mission.

The European Union Rule of Law Mission (EULEX) in Kosovo, which deployed in December 2008, is the largest civilian mission ever launched under the Common Security and Defence Policy (CSDP). The central aim is to assist and support the Kosovo authorities in the rule of law area, specifically in the police, judiciary and customs areas. EULEX works closely with KFOR in the field.

C) Cooperation in other regions

- Afghanistan

Over the past decade, NATO and the EU have played key roles in bringing peace and stability to Afghanistan, as part of the international community's broader efforts to implement a comprehensive approach to assist the country. The

NATO-led International Security Assistance Force (ISAF) helped create a stable and secure environment in which the Afghan government as well as other international actors could build democratic institutions, extend the rule of law and reconstruct the country.

Both ISAF and its successor Resolute Support Mission have cooperated with the EU's Rule of Law Mission (EUPOL), which operated in Afghanistan from June 2007 to December 2016. EUPOL Advisers at the Afghan Ministry of Interior and the Afghan National Police supported the reform of the ministry and the development of civilian policing.

The EU also initiated a program for justice reform and helped to fund civilian projects in NATO-run Provincial Reconstruction Teams that were led by an EU member country.

- Darfur

Both NATO and the EU supported the African Union's mission in Darfur, Sudan in particular with regard to airlift rotations.

- Piracy

For several years NATO's naval forces deployed under Operation Ocean Shield (2008-2016) and EU naval forces (Operation Atalanta) worked side by side with other actors, off the coast of Somalia for anti-piracy missions.

- Iraq

  Both NATO and the EU are increasing their presence in Iraq. The EU is focusing on the civilian security sector. NATO is helping build the capacities of the Iraqi defence and security structures, and is scaling up these efforts with the launch of a non-combat NATO Training Mission launched at the Brussels Summit in July 2018, at the request of the Government of Iraq. NATO's mission will complement the ongoing efforts of the Coalition to Defeat ISIS/Daesh and other international actors.

Other areas of cooperation:

- Political consultation

  The range of subjects discussed between NATO and the EU has expanded considerably over the past two years, particularly on security issues within the European space or its immediate vicinity.

  Since the crisis in Ukraine, both organisations have regularly exchanged views on their respective decisions, especially with regard to Russia, to ensure that their messages and actions complement each other. Consultations have also covered developments in the Western Balkans, Libya and the Middle East.

- Capabilities

  Together with operations, capability development is an area where cooperation is essential and where there is potential for further growth. The NATO-EU

Capability Group was established in May 2003 to ensure the coherence and mutual reinforcement of NATO and EU capability development efforts.

Following the creation, in July 2004, of the European Defence Agency (EDA) to coordinate work within the EU on the development of defence capabilities, armaments cooperation, acquisition and research, EDA experts contribute to the work of the Capability Group.

Among other issues, the Capability Group has addressed common capability shortfalls in areas such as countering improvised explosive devices and medical support. The Group is also playing an important role in ensuring transparency and complementarity between NATO's work on Smart Defence and the EU's Pooling and Sharing initiative.

- Terrorism and WMD proliferation

Both NATO and the EU are committed to combating terrorism and the proliferation of weapons of mass destruction (WMD). They have exchanged information on their activities in the field of protection of civilian populations against chemical, biological, radiological and nuclear attacks.

- Civil emergency planning

The two organisations also cooperate in the field of civil emergency planning by exchanging inventories of measures taken in this area.

Participation:

- With the enlargement of both organisations in 2004, followed by the accession of Bulgaria, Romania and Croatia to the EU, the two organisations have 21 member countries in common (after the UK left the EU due to BREXIT). Albania, Canada, Iceland, Montenegro, Norway, Turkey and the United States, which are members of NATO but not of the EU, participate in all NATO-EU meetings. So do Austria, Finland, Ireland, Sweden, and since 2008, Malta, which are members of the EU and of NATO's Partnership for Peace (PfP) program.

- Cyprus, which is not a PfP member and does not have a security agreement with NATO on the exchange of classified documents, does however participate in official NATO-EU meetings. This is a consequence of decisions taken by NATO in December 2002.

Framework for cooperation:

- An exchange of letters between the NATO Secretary General and the EU Presidency in January 2001 defined the scope of cooperation and modalities of consultation on security issues between the two organisations. Cooperation further developed with the signing of the NATO-EU Declaration on ESDP in December 2002 and the agreement, in March 2003, of a framework for cooperation.

- NATO-EU Declaration on ESDP: The NATO-EU Declaration on ESDP, agreed on 16 December 2002, reaffirmed the EU assured access to NATO's planning capabilities for its own military operations and reiterated the political principles of the strategic partnership: effective mutual consultation; equality and due regard for the decision-making autonomy of the EU and NATO; respect for the interests of EU and NATO member states; respect for the principles of the Charter of the United Nations; and coherent, transparent and mutually reinforcing development of the military capability requirements common to the two organisations.

- The "Berlin Plus" arrangements: As part of the framework for cooperation adopted on 17 March 2003, the so-called "Berlin Plus" arrangements provide the basis for NATO-EU cooperation in crisis management in the context of EU-led operations that make use of NATO's collective assets and capabilities, including command arrangements and assistance in operational planning. In effect, they allow the Alliance to support EU-led operations in which NATO as a whole is not engaged.

- NATO and the EU meet on a regular basis to discuss issues of common interest. Meetings take place at different levels including at the level of foreign ministers, ambassadors, military representatives and defence advisors. There are regular staff-to-staff talks at all levels between NATO's International Staff and International Military Staff, and their respective EU interlocutors (the European External Action Service, the European Defence Agency, the European Commission and the European Parliament).

- Permanent military liaison arrangements have been established to facilitate cooperation at the operational level. A NATO Permanent Liaison Team has been operating at the EU Military Staff since November 2005 and an EU Cell was set up at SHAPE (NATO's strategic command for operations in Mons, Belgium) in March 2006.

As a summary of the key points in the relations between NATO and the EU, we will point out:

a) Relations between NATO and the EU were institutionalised in the early 2000s, building on steps taken during the 1990s to promote greater European responsibility in defence matters (NATO-Western European Union cooperation).

b) The 2002 NATO-EU Declaration on a European Security and Defence Policy (ESDP) reaffirmed EU assured access to NATO's planning capabilities for the EU's own military operations.

c) In 2003, the so-called "Berlin Plus" arrangements set the basis for the Alliance to support EU-led operations in which NATO as a whole is not engaged.

d) At the 2010 Lisbon Summit, the Allies underlined their determination to improve the NATO-EU strategic partnership. The 2010 Strategic Concept committed the Alliance to working more closely with other international organisations to prevent crises, manage conflicts and stabilise post-conflict situations.

e) In Warsaw in July 2016, the two organisations outlined areas for strengthened cooperation in light of common challenges to the east and south, including countering hybrid threats, enhancing resilience, defence capacity building, cyber defence, maritime security and exercises. As a follow-up, in December 2016, NATO foreign ministers endorsed a statement to which were annexed 42 common measures to advance NATO-EU cooperation. A further 32 measures were agreed in December 2017.

f) On 10 July 2018, in a joint declaration, the two organisations agreed to focus on swift progress in the areas of military mobility, counter-terrorism and strengthening resilience to chemical, biological, radiological and nuclear-related risks as well as promoting the women, peace and security agenda.

g) Allied leaders welcomed this joint declaration at the Brussels Summit in July 2018 and tangible results achieved since 2016. They recognised that the development of European defence capabilities, while ensuring coherence and complementarity and avoiding unnecessary duplication, is key in joint efforts to make the Euro- Atlantic area safer and contributes to transatlantic burden-sharing

h) Close cooperation between NATO and the EU is an important element in the development of an international "comprehensive approach" to crisis management and operations, which requires the effective application of both military and civilian means.

i) NATO and the EU currently have 21 member countries in common (after the UK left the EU due to BREXIT).

# RELATIONS WITH OTHER INTERNATIONAL ORGANISATIONS

## A comprehensive approach to crises

Lessons learned from NATO operations show that addressing crisis situations calls for a comprehensive approach combining political, civilian and military instruments. Building on its unique capabilities and operational experience, including expertise in civilian-military interaction, NATO can contribute to the efforts of the international community for maintaining peace, security and stability, in full coordination with other actors. Military means, although essential, are not enough on their own to meet the many complex challenges to our security. The effective implementation of a comprehensive approach to crisis situations requires nations, international organisations and non-governmental organisations to contribute in a concerted effort.

Planning and conduct operations:

NATO takes full account of all military and non-military aspects of crisis management, and is working to improve practical cooperation at all levels with all relevant organisations and actors in the planning and conduct of operations. The Alliance promotes the clear definition of strategies and objectives among all relevant actors before launching an operation, as well as enhanced cooperative planning.

The Allies agree that, as a general rule, elements of stabilisation and reconstruction are best undertaken by those actors and organisations that have the relevant expertise, mandate and competence. However, there can be circumstances which may hamper other actors from undertaking these tasks, or undertaking them without support from NATO.

To improve NATO's contribution to a comprehensive approach of the international community when addressing crises, NATO bodies as well as individual Allies follow the Comprehensive Approach Action Plan to promote integrated civil-military planning across NATO's three core tasks (collective defence, crisis management and cooperative security). The principles of the comprehensive approach (coherence of actions, civil-military interaction and reaching out to external partners) are integral to the activities of the NATO Headquarters 'Crisis Management Task Force as well as the NATO Command and Force Structures.

The planning and conduct of NATO operations and missions now integrate perspectives from different priority areas including: gender; the Women, Peace and Security agenda; children and armed conflict; building integrity; cultural property protection; combating trafficking in human beings; and environmental protection.

Lessons learned, training, education and exercises:

Applying a comprehensive approach means a change of mindset. The Alliance therefore emphasises joint training of civilian and military personnel to promote the sharing of lessons learned and to build trust and confidence between NATO, its partners and other international and local actors.

In some cases, lessons learned are being developed at staff level, for example, with the United Nations, related to

Libya. Another example is the NATO Defence Education Enhancement Program, which, as a matter of principle, reaches out to external providers and enablers including international organisations and non-governmental organisations, addressing both civilian and military experts. And Sweden, a key NATO partner, designed its 2018 Viking exercise (involving 2,500 participants from 50 countries and 35 organisations) on comprehensive approach principles.

Enhancing cooperation with external actors:

Cooperation has become well established with the United Nations and its agencies, the European Union and the Organization for Security and Co-operation in Europe, in particular, as well as with the World Bank, the International Committee of the Red Cross, the International Organization for Migration, the African Union, INTERPOL and the League of Arab States. Closer links are developed with non-governmental organisations as well.

The scope of cooperation ranges from political dialogue, including through regular staff talks, "NATO education days" and the yearly Comprehensive Approach Awareness Course, to practical cooperation during operations and missions. High-level officials from these organisations are regularly invited to meetings of the North Atlantic Council, including at the level of heads of states and governments, to discuss closer cooperation and issues of common interest.

The implementation of the comprehensive approach has helped to build mutual awareness with these organisations. This has allowed the Alliance to broaden the range of its external interlocutors, who are becoming more accustomed to work with NATO and better informed about the role of military in complex environments.

Strategic communications:

To be effective, a comprehensive approach to crisis management must be complemented by sustained and coherent public messages. NATO's information campaigns are substantiated by systematic and updated information, documenting progress in relevant areas. Efforts are also being made to share communication strategies with international actors and to coordinate communications in theater.

Finally, we will highlight that:

1. Different actors contribute to a comprehensive approach based on a shared sense of responsibility, openness and determination, taking into account their respective strengths, mandates and roles, as well as their decision-making autonomy.

2. In December 2017, NATO reviewed the tasks of its 2011 Comprehensive Approach Action Plan, validating the importance of civilian-military interaction and cooperation with other actors.
3. These tasks are being implemented by a dedicated civilian-military task force that involves all relevant NATO bodies and commands.

4. The Action Plan covers four key areas: planning and conduct of operations; lessons learned, training, education and exercises; cooperation with external actors; and strategic communications.

5. Implementation of the comprehensive approach is integral to many recent and ongoing NATO activities, such as its contributions to the international community's fight against terrorism and efforts to project stability, and its role in responding to hybrid threats.


**NATO's relations with the United Nations**

Let us start by saying that the United Nations (UN) is at the core of the wider institutional framework within which the Alliance operates, a principle which is enshrined in NATO's founding treaty. UN Security Council resolutions have provided the mandate for NATO's major peace-support operations in the Balkans and in Afghanistan, and also provide the framework for NATO's training mission in Iraq. More recently, NATO has provided logistical assistance to the African Union's UN-endorsed peacekeeping operation in the Darfur region of Sudan.

Lately, cooperation between NATO and the United Nations has developed well beyond their common engagement in the western Balkans and in Afghanistan. The relationship between the two organisations has been steadily growing at all levels – on the ground, conceptually and politically, as well as institutionally. Cooperation and consultations with UN specialised bodies go beyond crisis management and cover a wide range of issues, including civil emergency planning, civil-military cooperation, combating human trafficking, action against mines, and the fight against terrorism.

The North Atlantic Treaty and the UN Charter:

It is important pointing that the acknowledgement of a direct relationship between the North Atlantic Treaty and the Charter of the United Nations is a fundamental principle of the Alliance. The Charter, signed in San Francisco on 26 June 1945 by fifty countries, provides the legal basis for the creation of NATO and establishes the overall responsibility of the UN Security Council for international peace and security. These two fundamental principles are enshrined in NATO's North Atlantic Treaty signed in Washington on 4 April 1949.

Thus, the preamble to the Washington Treaty makes it clear that the UN Charter is the framework within which the Alliance operates. In its opening phrases, the signatories of the Treaty reaffirm their faith in the purposes and principles of the Charter. In Article 1 they also undertake to settle international disputes by peaceful means and to refrain from the threat or use of force in any manner inconsistent with the purposes of the UN Charter. Article 5 of the Treaty makes explicit reference to Article 51 of the UN Charter in asserting the right of the Allies to take, individually or collectively, such action as they deem necessary for their self-defence. This includes the use of armed force. Moreover, it commits the member countries to terminating any such armed attack and all measures taken as a result, when the UN Security Council has itself taken the measures necessary to restore and maintain international peace and security.

Moreover, further reference to the UN Charter can be found in Article 7 of the North Atlantic Treaty, which states that the Treaty does not affect and shall not be interpreted as affecting in any way the rights and obligations of Allies under the Charter and reaffirms the primary responsibility of the UN Security Council for the maintenance of international peace and security. And finally, a clause in Article 12 of the Treaty provides for a review of the Treaty after ten years if any of the Parties to it so requests. It stipulates that the review would take place in the light of new developments affecting peace and security in the North Atlantic area, including the development of universal and regional arrangements under the UN Charter.

Finally, the North Atlantic Treaty came into force on 24 August 1949. None of the Parties to it have requested a review of the Treaty under Article 12, although at each stage of its development the Alliance has kept the implementation of the Treaty under continuous review for the purpose of securing its objectives.

Framework for cooperation:

In September 2008, building on the experience of over a decade of working together, the Secretaries General of the two organisations agreed to establish a framework for expanded consultation and cooperation.

Since the signing of the 2008 framework, cooperation has continued to develop in a practical way, taking into account each organisation's specific mandate, expertise, procedures and capabilities. Regular exchanges and dialogue at senior and working levels on political and operational issues have become a standard feature of the inter-institutional relationship. NATO's Secretary General reports regularly to the UN Secretary-General on progress in UN-mandated NATO-led operations and on other key decisions of the

North Atlantic Council, including in the area of crisis management and in the fight against terrorism. The UN is frequently invited to attend NATO ministerial meetings and summits; the NATO Secretary General participates in the UN General Assembly; and staff level meetings, covering the broad range of cooperation and dialogue, take place on an annual basis between the secretaries of NATO and the UN.

Key areas of cooperation:

A) Peace operations

- NATO's unique capabilities and experience can be a valuable source of support to the UN, whose peacekeepers operate in increasingly challenging and dangerous environments. NATO and UN staffs have worked to build practical cooperation in this domain.

- At the 2015 Leaders' Summit on Peacekeeping, the NATO Secretary General pledged to enhance support to the UN, in particular in the areas of countering improvised explosive devices, training and preparedness, supporting the UN's efforts to deploy more rapidly and working more closely on capacity building in countries at risk, both with the UN Nations and the EU. As the UN reforms its approach to peace operations, NATO will continue to look for where its support can make a difference.

B) Counter-terrorism

- The UN Global Counter-Terrorism Strategy, international conventions and protocols against terrorism, together with relevant UN Security Council Resolutions provide the framework for NATO's efforts to combat terrorism. We want to emphasise that NATO works closely at staff and committee level with the UN Counter-Terrorism Committee (UN CTC) and its Executive Directorate, as well as with the Counter-Terrorism Implementation Task Force and many of its component organisations. The Terrorism Prevention Branch of the UN Organization for Drugs and Crime is also an important partner for NATO.

C)  Non-proliferation

- NATO contributes to the work of the UN Security Council Committee established following the adoption of UNSCR 1540 (2004), which addresses the threat to international peace and security posed by the proliferation of nuclear, chemical and biological weapons and their means of delivery.

- In this context, since 2004 the Alliance has been organising a string of international non-proliferation conferences and seminars with the active participation of partner countries and international organisations, latest in Ljubljana, Slovenia on 9 and 10 May 2016.

- NATO has also addressed the implementation of UNSCR 1540 at regional and sub-regional levels, including through its Science for Peace and Security Program, and will continue to address the need for assistance of partner countries upon request.

D) Women, Peace and Security

- NATO remains committed to the full implementation of UNSCR 1325 on Women, Peace and Security and related Resolutions, which aim to protect and promote women's rights, role and participation in preventing and ending conflict. In line with the policy developed by NATO Allies together with partners in the Euro-Atlantic Partnership Council (EAPC), significant progress has been made in implementing the goals set out in these Resolutions.

- In this regard, NATO has endorsed a Strategic Report on mainstreaming UNSCR 1325 and related Resolutions across NATO's core activities: collective defence; crisis management and operations; and cooperative security. An updated NATO Action Plan for the Implementation of the UNSCR 1325/EAPC Policy on Women, Peace and Security was also agreed.

- In October 2015, NATO's Deputy Secretary General participated in the UN Security Council Open Debate on Women, Peace and Security, and pledged to do more in this area, including sharing best practices and lessons learned on increasing female participation at decision-making levels with Allies and partners; encouraging Allies to submit female candidates for NATO's most senior decision-making positions; strengthening partnerships with international organisations like the UN, OSCE, the EU and the African Union on gender equality, as well as institutionalising the engagement of civil society in the development, execution and monitoring of the NATO/EAPC Action Plan on Women, Peace, and Security.

- NATO also committed to financing evidence-based research aimed at understanding the role of gender in preventing and countering violent extremism, which complimented the adoption in 2015 of UNSCR 2242 at the 15th anniversary commemoration of UNSCR 1325.

E) Protecting children in armed conflict

- NATO is committed to the implementation of UNSCR 1612 and related Resolutions on the protection of children affected by armed conflict. At the 2014 NATO Summit in Wales, NATO leaders decided more could be done to ensure the Alliance is sufficiently prepared whenever and wherever the issue of Children and Armed Conflict is likely to be encountered. The result was the NATO policy document "The Protection of Children in Armed Conflict - Way forward".

- Prepared in cooperation with the UN, the policy aims to deepen the implementation of the UNSCR 1612 into NATO operations and missions. These efforts include training the Alliance's deployed troops to recognise, monitor and report violations against children and to incorporate child protection issues into NATO exercise scenarios. When it is invited to train local forces, NATO also emphasises the importance of protecting children in armed conflict. NATO also recently appointed to Children and

Armed Conflict Advisor as part of the Resolute Support Mission in Afghanistan.

F) Small arms and light weapons

- NATO supports the implementation of the UN Program of Action to Prevent, Combat and Eradicate the Illicit Trade in Small Arms and Light Weapons (SALW), adopted in July 2001 by nearly 150 countries, including all NATO member states. The Alliance also participates in UN experts' meetings and review conferences. The NATO/EAPC Ad Hoc Working Group on SALW and Mine Action as well as the Trust Fund mechanism were established in 1999 to support partner countries in implementing provisions of the Ottawa Convention (also known as the Antipersonnel Mine Ban Convention).

- Moreover, NATO supports nations in implementing the Arms Trade Treaty that entered into force in December 2014 through training and application of standards.

- NATO has also worked closely with UN agencies to develop international standards for ammunition life-cycle management, such as the International Ammunition Technical Guidelines. The Alliance also strives to support regional and sub-regional efforts with the UN and partners beyond the EAPC area in managing SALW, ammunition, explosive remnants of war. In this context, NATO developed some capacities through its Science for Peace and Security Program.

G) Disaster relief

- NATO also cooperates with the UN in support of disaster-relief operations. Through the Euro-Atlantic Disaster Response Coordination Center (EADRCC), NATO coordinates consequence-management efforts with UN and other bodies and shares information on disaster assistance. All the EADRCC's tasks are performed in close cooperation with the UN Office for the Coordination of Humanitarian Affairs (UN OCHA), which retains the primary role in the coordination of international disaster-relief operations.

- The EADRCC is a regional coordination mechanism, supporting and complementing the UN efforts. In the case of a disaster requiring international assistance, it is up to individual NATO member and partner countries to decide whether to provide assistance, based on information received from the EADRCC.

Evolution NATO-UN cooperation in the field:

Working relations between the United Nations and the Alliance were limited during the Cold War. This changed in 1992, against the background of growing conflict in the Western Balkans, where their respective roles in crisis management led to an intensification of practical cooperation in the field.

1) Bringing peace to the former Yugoslavia

- In July 1992, NATO ships belonging to the Alliance's Standing Naval Force Mediterranean, assisted by NATO maritime patrol aircraft, began monitoring operations in the Adriatic in support of a UN arms embargo against all republics of the former Yugoslavia. A few months later, in November 1992, NATO and the Western European Union (WEU) began enforcement operations in support of UN Security Council Resolutions aimed at preventing the escalation of the conflict.

- The readiness of the Alliance to support peacekeeping operations under the authority of the UN Security Council was formally stated by NATO foreign ministers in December 1992. A number of measures were subsequently taken, including joint maritime operations under the authority of the NATO and WEU Councils: NATO air operations; close air support for the United Nations Protection Force (UNPROFOR); air strikes to protect A "Safe Areas"; and contingency planning for other options which the UN might take.

- Following the signature of the General Framework Agreement for Peace in Bosnia and Herzegovina (the Dayton Agreement) on 14 December 1995, NATO was given a mandate by the United Nations, on the basis of UN Security Council Resolution 1031, to implement the military aspects of the peace agreement. NATO's first peacekeeping operation - the Implementation Force (IFOR)- began operations in Bosnia and Herzegovina to fulfil this mandate in December 1995. One year later, it was replaced by the NATO-led Stabilisation Force (SFOR). Throughout their mandates both multinational forces worked closely with other international organisations and humanitarian agencies on the ground, including UN agencies such as the UN High Commissioner for Refugees (UNHCR) and the UN International Police Task Force (IPTF).

- From the onset of the conflict in Kosovo in 1998 and throughout the crisis, close contacts were maintained between the UN Secretary-General and NATO's Secretary General. Actions were taken by the Alliance in support of UN Security Council Resolutions both during and after the conflict. The Kosovo Force (KFOR) was deployed on the basis of UN Security Council Resolution 1244 of 12 June 1999 to provide an international security presence as the prerequisite for peace and reconstruction of Kosovo. Throughout its deployment, KFOR has worked closely with the UN Interim Administration Mission in Kosovo (UNMIK) as well as with other international and local stakeholders.

- In 2000 and 2001, NATO and the United Nations also cooperated successfully in containing major ethnic discord in southern Serbia and preventing a full-blown civil war in the Republic of North Macedonia (then known as the former Yugoslav Republic of Macedonia).

2) Afghanistan

- Cooperation between NATO and the UN is playing a key role in Afghanistan. The Alliance formally took over the International Security Assistance Force (ISAF), a UN-mandated force, in August 2003. Originally tasked with helping to provide security in and around Kabul, ISAF was subsequently authorised by a series of UN Security Council Resolutions to expand its presence into other regions of the country to extend the authority of the central government and to facilitate development and reconstruction.

- NATO and ISAF worked closely with the United Nations Assistance Mission in Afghanistan (UNAMA) and other international actors that are supporting governance, reconstruction and development. The close cooperation took place in various settings, in Afghanistan as well as in UN and NATO capitals. It included co-membership of the Joint Coordination and Monitoring Board (JCMB) overseeing the implementation of the internationally endorsed Afghanistan Compact, co-chairmanship together with the Afghan Government of the Executive Steering Committee for Provincial Reconstruction Teams, and other joint Afghan-international community bodies.

- NATO and the UN continue consulting closely on their respective postures in Afghanistan. NATO is keeping the UN well informed on the Resolute Support Mission.

3) Iraq

- Under the terms of UN Security Council Resolution 1546 and at the request of the Iraqi Interim Government, NATO provided assistance in training and equipping Iraqi security forces through the

NATO Training Mission-Iraq (NTM-I) from 2004 to end 2011.

4) Supporting African Union missions

- In June 2005, following a request from the African Union (AU) and in close coordination with the United Nations and the European Union, NATO agreed to support the AU's Mission in Sudan (AMIS), which is trying to end the continuing violence in the Darfur region. NATO assisted by airlifting peacekeepers from African troop-contributing countries to the region and also helped train AU troops in how to run a multinational military headquarters and how to manage intelligence.

- Following a request from the AU in 2007, NATO accepted to assist the AU Mission in Somalia (AMISOM) by providing airlift support to AU member states willing to deploy on this mission. NATO is also providing capacity-building assistance for the AU via a Senior Military Liaison Office in Addis Ababa, Ethiopia.

5) Deterring piracy

- In October 2008, NATO agreed to a request from the UN Secretary-General to deploy ships off the coast of Somalia to deter piracy and escort merchant ships carrying World Food Program cargo.

6) Libya

- On 27 March 2011, NATO Allies decided to take on the whole military operation in Libya under United Nations Security Council Resolution 1973. The purpose of Operation Unified Protector was to protect civilians and civilian-populated areas under threat of attack. NATO implemented all military aspects of the UN Security Council Resolution. Allies moved swiftly and decisively to enforce the arms embargo and no-fly zone called for in the resolution, and to take further measures to protect civilians and civilian-populated areas from attack. Operation Unified Protector was concluded on 31 October 2011.

To summarise we will say that:

1. NATO's 2010 Strategic Concept commits the Alliance to preventing crises, managing conflicts and stabilising post-conflict situations, including by working more closely with NATO's international partners, most importantly the UN and the European Union (EU).

2. UN Security Council Resolutions have provided the mandate for NATO's operations in the Western Balkans, Afghanistan and Libya. They have also provided the framework for NATO's training mission in Iraq.

3. NATO has also provided support to UN-sponsored operations, including logistical assistance to the African Union's UN-endorsed peacekeeping operations in Darfur, Sudan, and in Somalia; support for UN disaster-relief operations in Pakistan,

following the massive earthquake in 2005; and escorting merchant ships carrying World Food Program humanitarian supplies off the coast of Somalia.

4. Practical cooperation between NATO and the UN extends beyond operations to include: crisis assessment and management, civil-military cooperation, training and education, tackling corruption in the defence sector, mine action, mitigating the threat posed by improvised explosive devices, civilian capabilities, promoting the role of women in peace and security, the protection of civilians, including children, in armed conflict, combating sexual and gender-based violence, arms control and non-proliferation, and the fight against terrorism.

5. At the 2015 Leaders 'Summit on Peacekeeping, held on the margins of the 70th UN General Assembly, NATO also pledged enhanced support to the UN in the area of peace operations.

6. An updated Joint Declaration setting out plans for future cooperation between NATO and the UN was signed on 26 October 2018. Building on the original Joint Declaration, signed in September 2008, it sets out priority areas for future cooperation, including support to UN peace operations, countering terrorism, the protection of civilians, and promoting the Women, Peace and Security agenda.

7. In 2010, NATO reinforced its liaison arrangements by establishing the post of NATO Civilian Liaison

Officer to the United Nations, in addition to that of a Military Liaison Officer, established in 1999.

8. Enhanced cooperation with the UN - and other international actors such as the EU and the Organization for Security and Co-operation in Europe - is an integral part of NATO's contribution to a "Comprehensive Approach" to crisis management and operations.

## NATO and the Organization for Security and Co-operation in Europe

The Organization for Security and Co-Operation in Europe (OSCE) is an important partner for NATO. Allies attach great importance to the role of the OSCE in fostering dialogue, building trust, and upholding the rules-based international order. The OSCE establishes the principles that govern international relations in the Euro-Atlantic area and embodies a comprehensive approach to human security. The two organisations play complementary roles in building security and maintaining stability in the Euro-Atlantic area. Both support the principles that underpin the European security order. Both also acknowledge the need for a coherent and comprehensive approach to crisis management, which requires effective application of both military and civilian means.

The political basis for cooperation:

It is basic to know that the NATO-OSCE relationship reflects the Alliance's commitment to a broad approach to security and the desire of NATO member countries, expressed in the Alliance's 1999 Strategic Concept, to establish cooperative relationships with other complementary and mutually reinforcing organisations.

So, political relations between the two organisations are governed by the "Platform for Cooperative Security" launched at the OSCE Istanbul Summit in 1999. The Platform calls for reinforced cooperation between international organisations, drawing on the resources of the international community to drive for democracy, prosperity and stability in Europe and beyond. It provides for meetings between the organisations to discuss operational and political issues of common interest.

In November 2002, at the Prague Summit, NATO leaders expressed their desire to extend cooperation with the Organization for Security and Cooperation in Europe in the areas of conflict prevention, crisis management and post-conflict rehabilitation operations. They also highlighted the need to exploit the complementarity of international efforts aimed at reinforcing stability in the Mediterranean region. Following this statement, the two organisations began developing closer contacts affecting their respective dialogues with countries in the region.

Otherwise, in light of changes in the security environment, both organisations have also extended their dialogue to other areas of common interest, including terrorism. In December 2003, the OSCE Ministerial Council, meeting at Maastricht, the Netherlands, adopted a new "Strategy to Address Threats to Security and Stability in the 21st Century". This document recalls the need, in a constantly changing security environment, to interact with other organisations and institutions cooperating in the context of the Platform for

Cooperative Security, and to take advantage of the assets and strengths of each.

Besides, with regard to the fight against terrorism, NATO's efforts, particularly within the framework of partnerships with non-member countries, complement those of the Organization for Security and Co-operation in Europe. A number of OSCE initiatives have been launched since the September 2001 terrorist attacks on the United States, including the Charter on Preventing and Combating Terrorism adopted in Porto in 2002.

Practical cooperation:

It should be remembered that practical NATO-OSCE cooperation is best exemplified by the complementary missions undertaken by the two organisations in the western Balkans. In 1996, after the signing of the Dayton Peace Accord they developed a joint action programme in Bosnia and Herzegovina. The NATO-led Implementation Force (IFOR) established to implement the military aspects of the peace agreements, and the Stabilisation Force (SFOR) which succeeded it, provided vital support for implementation of the civilian aspects of the agreements. By ensuring the security of OSCE personnel and humanitarian assistance, NATO contributed, among others, to the smooth organisation of elections in Bosnia and Herzegovina under OSCE auspices.

The Organization for Security and Co-operation in Europe, in October 1998, established a Kosovo Verification Mission to monitor compliance on the ground with cease-fire agreements concluded after the deterioration of the situation in Kosovo and the efforts of the international community to avert further conflict. NATO conducted a parallel aerial surveillance mission. Following a further deterioration in security conditions, the OSCE Verification Mission was forced to withdraw in March 1999.

Then, following the NATO air campaign in Kosovo, in July 1999, a new OSCE Mission to Kosovo was established as part of the United Nations Interim Administration Mission in Kosovo. The role of the OSCE Mission, among other things, is to oversee the progress of democratisation, the creation of institutions and the protection of human rights. The Mission maintains close relations with the NATO-led Kosovo Force (KFOR), which has a mandate from the United Nations to guarantee a safe environment for the work of the international community.

Thus, NATO has also cooperated closely with the Organization for Security and Co-operation in Europe in the former Yugoslav Republic of Macedonia.

In September 2001, a NATO task force was set up to provide additional security for EU and OSCE observers monitoring the implementation of a framework peace agreement, which had been reached in the summer after a period of internal ethnic unrest in the spring. The European Union officially took over this operation, renamed Concordia, from March 2003 until its conclusion in December 2003.

In this way, NATO-OSCE cooperation has also contributed to promoting better management and securing of borders in the western Balkans. At a highlevel conference held in Ohrid in May 2003, five Balkan countries endorsed a Common Platform developed by the European Union,

NATO, the Organization for Security and Co-operation in Europe and the Stability Pact, aimed at enhancing border security in the region. Each organisation supports the countries involved in the areas within its jurisdiction.

Further, NATO and the Organization for Security and Co-operation in Europe also seek to coordinate their efforts in other areas. Initiatives taken by NATO in areas such as arms control, mine clearance, elimination of ammunition stocks and efforts to control the spread of small arms and light weapons dovetail with OSCE efforts aimed at preventing conflict and restoring stability after a conflict. Moreover, in the regional context, both organisations place special emphasis on southeastern Europe, the Caucasus and Central Asia. Each has also developed parallel initiatives directed towards the countries of the Mediterranean region.

Regular contacts:

Equally, contacts between NATO and the Organization for Security and Cooperation in Europe take place at different levels and in different contexts, including high-level meetings between the NATO Secretary General and the OSCE Chairman-in-Office. Periodically, NATO's Secretary General is invited to address the OSCE Permanent Council. Similarly, the North Atlantic Council may from time to time invite the OSCE Chairman-in-Office to address one of its meetings.

Thus, exchanges of views on issues of common interest such as crisis management, border security, disarmament and terrorism regularly take place between officials from both organisations. Their respective representatives in the field meet regularly to share information and discuss various aspects of their cooperation.

As a summary of the cooperation between both structures, we will fix the following points:

1. NATO and the OSCE cooperate at both the political and operational levels in a range of areas including: conflict prevention and resolution; post-conflict rehabilitation including border security; countering the proliferation of small arms and light weapons, and arms control; promoting the Women, Peace and Security agenda; counter-terrorism; and addressing emerging security challenges.

2. At the political level, NATO and the OSCE exchange views on thematic and regional security issues of common interest through exchanges by senior leadership, direct cooperation and regular staff-to-staff talks.

3. The two organisations complement each other's efforts in the field: NATO initiatives to support defence reform, mine clearance and the destruction of stockpiles of arms and munitions, dovetail with OSCE efforts aiming to build peace and stability (successful examples of such cooperation include Central Asia, the Western Balkans and South Caucasus).

4. At recent summits, the Allies have reiterated the importance of the OSCE's role in regional security and as a forum for dialogue on issues relevant to

Euro-Atlantic security, not least on arms control and disarmament.

5.  NATO Allies fully support the promotion of arms control, military transparency, and confidence- and security-building measures through the modernisation of all of the political-military tools in the OSCE toolbox, especially the Vienna Document.

6.  The Alliance aims to further enhance NATO's cooperation with the OSCE as decided at the Warsaw Summit. A permanent liaison presence has been established to that effect in Vienna.

## Cooperation with other international organisations

We will comment that NATO is keen to deepen its relations with other international organisations to share information and promote appropriate and effective action in areas of common interest. The primary focus of its relations with other international organisations concerns cooperation with the European Union, the United Nations and the Organization for Security and Co-operation in Europe, as described in the previous chapters. NATO also holds consultations and engages in differing forms of cooperation with a number of other important international institutions.

1)  The Council of Europe:

On 5 May 1949, the Council of Europe was established "to achieve a greater unity between its members for the purpose of safeguarding and realising the ideals and principles which are their common heritage and facilitating their social and economic progress".

The Council's overall aim is to maintain the basic principles of humanitarian rights, pluralist democracy and the rule of law, and to enhance the quality of life of European citizens. NATO regularly receives documents, reports and records from the Council of Europe and is kept informed of different parliamentary sessions or upcoming events. The outcome of various sessions and reports on issues of common interest is monitored by NATO's International Staff and this information is distributed to relevant divisions within the organisation.

2) The International Organization for Migration:

It is interesting to know that the International Organization for Migration is the leading international organisation working with migrant populations and governments on issues relating to migration challenges. It is committed to the principle that humane and orderly migration benefits both migrants and the societies in which they live. Established in 1951 and tasked with the resettlement of European displaced persons, refugees and migrants, the organisation now encompasses a variety of migration-management issues and other activities throughout the world.

Likewise, with offices and operations on every continent, the organisation helps governments and civil societies, for example, in responding to sudden migration flows, post-emergency returns and reintegration programmes, and providing assistance to migrants on their way to new homes.

It also promotes the training of officials and measures to counter trafficking in human beings.

This way, cooperation with NATO takes place in several fields such as combating trafficking in human beings, border security and reconstruction in post-conflict regions. Regions where there is great potential for cooperation include the Caucasus and Central Asia. The first formal and structured contacts between the two organisations took place in staff-level meetings in September 2004.

3)  The Assembly of the Western European Union:

As we know, NATO also had contacts with the Assembly of the Western European Union (WEU). Although not an international organisation in the strict sense of the term, the Assembly was created in 1954 under the modified Brussels Treaty of 1948, which is the founding document of the Western European Union. Called upon in 1984 to contribute to the process of establishing a stronger European security and defence identity, the Western European Union was later relieved of these responsibilities, which were transferred to the European Union at the end of 1999 in the context of the latter's evolving European Security and Defence Policy. The Western European Union itself remained extant with a small secretariat located in Brussels with residual responsibilities.

Thus, the WEU Assembly remained active as an inter-parliamentary forum for general strategic reflection and contributes to intergovernmental and public debate on security and defence matters. National parliamentarians from 28 European countries sent delegations to the Assembly, which currently had 370 members. Its work was allocated to four principal committees dealing respectively with defence matters, political issues, matters relating to technology and aerospace, and parliamentary and public relations. The WEU Assembly met at least twice a year for plenary sessions and throughout the year in committee meetings, conferences and colloquia.

Recall again that at the turn of the 21st century, after the end of the Cold War, WEU tasks and institutions were gradually transferred to the European Union (EU), providing central parts of the EU's new military component, the European Security and Defence Policy (ESDP). This process was completed in 2009 when a solidarity clause between the member states of the European Union, which was similar (but not identical) to the WEU's mutual defence clause, entered into force with the Treaty of Lisbon. The states party to the Modified Treaty of Brussels consequently decided to terminate that treaty on 31 March 2010. On 30 June 2011, the WEU was officially declared defuncted.

4) The Organization for the Prohibition of Chemical Weapons:

Finally, another of the organisations with which NATO cooperates in the field of civil emergency planning is the Organization for the Prohibition of Chemical Weapons.

The Organization, established in 1997 by the countries that joined the Chemical Weapons Convention, seeks to ensure that the Convention works effectively and achieves its purpose. All NATO Allies are members of the Organization, which currently totals 174 member states.

Let us emphasise that one of the Organization's responsibilities is to provide assistance and protection to countries if they are attacked or threatened with chemical weapons, including by terrorists. It is in this area in particular that the Organization can be helpful to NATO's civil protection efforts which, following the September 2001 terrorist attacks on the United States, have increasingly focused on protecting populations against the potential consequences of attacks using chemical, biological, radiological and nuclear agents.

5) The International Committee of the Red Cross:

Let us highlight that one of the most significant non-governmental organisations with which NATO cooperates is the International Committee of the Red Cross: an impartial, neutral and independent organisation exclusively concerned with humanitarian action to protect the lives and dignity of victims of war and internal violence and to provide them with necessary assistance.

The International Committee of the Red Cross directs and coordinates international relief activities conducted in situations of conflict. In addition, the Committee endeavors to prevent suffering by promoting the strengthening of international humanitarian law and of universal humanitarian principles. It is in this context that its contacts with NATO have developed. In their operational planning, NATO authorities take account of the provisions of international humanitarian law; the operational plans embody references to implementation of and respect for international humanitarian law. While the dissemination of information on international humanitarian law is, in principle, a matter for the member states themselves, NATO may consider appropriate action to stimulate this process with the assistance of the International Committee of the Red Cross.

Moreover, relations between the two organisations have focused on ad hoc cooperation, with occasional informal exchanges of views between staff and high level meetings when required. Cooperation has taken place in the context of a number of issues in different countries and regions, for example in the Balkans, in Afghanistan and in Iraq.

Otherwise, the principal concerns of the International Committee of the Red Cross in the context of its relations with NATO relate to the application of international humanitarian law in armed conflicts, the complementarity of the military, political and humanitarian approaches to a crisis situation and respect for the differences between them, and NATO's responsibility regarding the implementation of international humanitarian law.

And at the practical level, the International Committee of the Red Cross has provided support for training courses on peacekeeping and civil emergency planning at the NATO School in Oberammergau, organised in the framework of the Partnership for Peace program. Cooperation has also taken place in the context of NATO's role with respect to Kosovo, and the Committee has provided input in the framework of some of the "lessons learned" evaluations undertaken by NATO.

Finally, in the context of civil emergency planning activities and exercises, NATO also often cooperates with the International Federation of the Red Cross and Red Crescent Societies.

**NATO-Parliamentary Assembly**

Let us start by saying that the NATO-Parliamentary Assembly (NATO PA) is an inter-parliamentary organisation which, since its creation in 1955, has acted as a forum for legislators from the North American and western European member countries of the Alliance to meet together to consider issues of common interest and concern.

While its principal objective is to foster mutual understanding among Alliance parliamentarians of the key security challenges facing the transatlantic partnership, its discussions also contribute to the development of the consensus among member countries that underpins the decision-making process in the Alliance.

So, the NATO Parliamentary Assembly is an inter-parliamentary organisation, which brings together legislators from NATO member countries to consider security-related issues of common interest and concern.

Since the 1980s, it has assumed additional roles by integrating into its work parliamentarians from NATO partner countries in Europe and beyond.

Fostering mutual understanding:

The Assembly's principal objective is to foster mutual understanding among Alliance parliamentarians of the key security challenges facing the transatlantic partnership. It is completely independent of NATO but provides a link between NATO and the parliaments of its member countries.

Work with parliamentarians from member countries:

- fostering dialogue among parliamentarians on major security issues;

- facilitating parliamentary awareness and understanding of key security issues and Alliance policies;

- providing NATO and its member governments with an indication of collective parliamentary opinion;

- providing greater transparency in NATO policies as well as collective accountability;

- strengthening the transatlantic relationship.

In fulfilling its goals, the Assembly provides a central source of information and a point of contact for member legislators and their respective national parliaments.

And let us highlight the cooperation with parliamentarians in partner countries since 1989:

- to assist in the development of parliamentary democracy throughout the Euro-Atlantic area by integrating parliamentarians from non-member countries into the Assembly's work;

- to assist parliaments of countries actively seeking Alliance membership;

- to increase cooperation with countries which seek closer relations with NATO rather than membership, including those of the Caucasus and the Mediterranean regions;

- to assist in the development of parliamentary mechanisms, practices and know-how essential for the effective democratic control of armed forces.

The NATO-Parliamentary Assembly is made up of 266 delegates from the 30 NATO member countries. Each delegation is based on the country's size and reflects the political composition of the parliament, therefore representing a broad spectrum of political opinion. Delegates are nominated by their parliaments according to their national procedures.

In addition to these NATO country delegates, delegates from 12 associate countries, four Mediterranean associate countries, as well as observers from eight other countries take part in its activities.

Inter-parliamentary assemblies such as the Organization for Security and Co-operation in Europe Parliamentary Assembly, the Council of Europe Parliamentary Assembly and the Western European Union Assembly also send delegations.

The European Parliament is entitled to send 10 delegates to Assembly Sessions and can participate in most committee and sub-committee activities.

Most of the Assembly's work is carried out by its five committees: the Committee on the Civil Dimension of Security; the Defence and Security Committee; the

Economics and Security Committee; the Political Committee; and the Science and Technology Committee.

There are several sub-committees, which meet during the year on fact-finding missions designed to gather information for sub-committee and committee reports. Sub-committee reports, like those produced directly for the committees, are amended and adopted by majority vote in the committees. Each year, the NATO-Parliamentary Assembly typically holds approximately 40 activities. These include two Plenary Sessions, a Standing Committee meeting, three to four Rose-Roth Seminars, two Mediterranean Seminars, 16 sub-committee meetings and a variety of other meetings.

The NATO-Parliamentary Assembly is headed by a President, who is a parliamentarian from a NATO member country. The headquarters of the Assembly comprises an International Secretariat of approximately 30 people based in Brussels, Belgium and is overseen by a Secretary General.

The International Secretariat performs a dual function: on the one hand, it conducts much of the research and analysis necessary for the substantive output of the Assembly's committees, and on the other hand, it provides the administrative support required to organise sessions, seminars, committee meetings, and other Assembly activities.

In addition, the International Secretariat maintains a close working relationship with NATO, other international organisations and research institutes. It also provides briefings on NATO-Parliamentary Assembly activities and concerns to visiting parliamentary groups, journalists, and academics.

**Atlantic Treaty Association and Youth Atlantic Treaty Association**

The Atlantic Treaty Association (ATA) is an independent organisation designed to support the values enshrined in the North Atlantic Treaty. Created on 18 June 1954, it is an umbrella organisation for the separate national associations, voluntary organisations and non-governmental organisations that formed to uphold the values of the Alliance after its creation in 1949. The Youth Atlantic Treaty Association (YATA) is the youth branch of the ATA and was formed in 1996.

The role of the ATA:

The ATA is a community of policy-makers, think tankers, diplomats, academics and representatives from industry. It seeks to inform the public of NATO's role in international peace and security and promote democracy, individual liberty and the rule of law through debate and dialogue.

To achieve this goal, it holds international seminars and conferences and launches initiatives, such as the Central and South Eastern European Security Forum and the Ukrainian Dialogue and Crisis Management Simulations.

The ATA is also active in NATO's Partnership for Peace (PfP) program, the Mediterranean Dialogue and the Istanbul Cooperation Initiative, launching conferences, seminars and multi-year research programs. As a result, the ATA's geographical scope has increased since the end of the Cold War, i.e. since the early 1990s, mirroring NATO's enlargement and its engagement with an ever-broader number of partner countries in the Euro-Atlantic area and beyond.

The ATA also cooperates with various organisations connected with Euro-Atlantic security, such as member associations of the ATA, the governments of member associations, the European Union, NATO and the NATO Parliamentary Assembly. It also promotes the development of civil society in, for instance, the Black Sea and Caucasus regions, and engages in dialogue with Middle Eastern countries.

More generally, the ATA fosters debate and dialogue in an effort to create a solid understanding of Alliance issues and current security issues such as hybrid warfare, cyber security and terrorism. In addition, it works to develop relations between organisations in different countries by connecting with civil society groups that support the basic principles of the North Atlantic Treaty. Furthermore, it seeks to develop relations between its members in an effort to achieve common goals.

The role of the YATA:

The ATA's youth division - YATA - was formed in 1996 during the ATA's General Assembly in Rome with the aim of reaching out to younger or "successor" generations.

It serves to bring together groups of young professionals working in security and defence, providing an opportunity for networking between themselves and senior level officials from different countries. It works in close cooperation with the ATA, supports its activities and shares its primary goals. They include educating and informing the successor generation about issues concerning international security, supporting research into NATO's role in the world and encouraging young leaders to shape the future of the transatlantic security relationship while promoting its importance.

The YATA also seeks to encourage cooperation between the youths of NATO member countries and partner countries, and between various international organisations to generate debate about the role of security institutions.

Although the YATA is officially part of the ATA, it also holds separate activities to achieve its objectives, such as its annual Atlantic Youth Seminars in Denmark (DAYS) and Portugal (PAYS), as well as crisis management simulations and regional conferences. The YATA also works with NATO's Public Diplomacy Division to organise international conferences and seminars where the national YATA chapters are able to meet Alliance leaders and officials, including the NATO Secretary General, to discuss transatlantic security issues.

Working mechanisms:

1) Structure:

- The ATA is composed of three main bodies: the Assembly, the Bureau and the Council, as well as the YATA and the Committee of Patrons.

- The Assembly is the top decision-making body of the ATA and is comprised of delegates from Member, Associate Member and Observer Member associations. With the exception of Observer Members, each delegate has one vote and resolutions are passed by a simple majority. In addition to the delegates, members of the press and academic community, government and military officials, and international observers may attend the General Assembly meetings, which are held once a year.

- The Bureau includes the president, vice presidents, secretary general, treasurer, YATA president and the legal adviser. Members of the Bureau assist in carrying out the decisions of the Council and the Assembly and aid in policy matters, in addition to developing relationships with other groups such as the NATO Parliamentary Assembly.

- The Council comprises Bureau members plus up to three delegates from each of the ATA Member, Associate Member and Observer Member associations. The ATA allows the Council to take action on its behalf, with the recommendation of the Bureau and the approval of the Assembly. The Council holds two meetings a year: once at NATO Headquarters and once in a host country.

2) The YATA:

- The Youth Atlantic Treaty Association is officially part of the ATA. It serves as the youth division of the ATA and has its own structure, activities and programs. Similarly to the ATA, there are separate national youth divisions.

3) The Committee of Patrons:

- The Committee of Patrons is comprised of previous ATA presidents and other people who have served the ATA with merit.

4) Officers:

- The President of the ATA is in charge of the general policy of the Association, in addition to acting as its spokesperson. The Assembly, with input from the Council, elects the president for a three-year period.

- The ATA Secretary General is in charge of day-to-day operations for the Association, furthering its goals and aims, implementing the decisions of the Assembly, Council and Bureau, and maintaining relationships with various other institutions. The Assembly, with input from the Council and the Bureau, elects the Secretary General for a three-year renewable period.

- The Assembly also elects the treasurer, who is in charge of financial matters, for a renewable three-year period.

5) Membership:

- There are three different types of membership in the ATA: Members, Associate Members and Observers.

6) Members:

- The national associations, which come from NATO member countries, may join the ATA as Members. As such, they may attend and participate in Bureau, Council and Assembly meetings. They also have full voting rights.

7) Associate Members:

- The national associations that make up the Associate Members of ATA come from non-NATO countries that have signed up to NATO's PfP program. Associate Members may attend and participate in Bureau, Council and Assembly meetings. Once an association's respective country joins NATO, the association automatically becomes a Member. Much like Members, Associate Members also have full voting rights.

8) Observer Members:

- Associations from non-NATO countries that have a direct interest in Euro-Atlantic security issues can participate in the ATA under the status of Observer Members. As Observer Members, the national associations may attend and participate in Council and Assembly meetings, but not Bureau meetings. Also, unlike Members and Associate Members, Observer Members have no voting rights.

The evolution of the ATA:

Following the creation of the Alliance in 1949, several separate organisations in NATO member countries formed with the aim of informing the public of NATO's role and activities. A few years later, these organisations came together under the umbrella of the Atlantic Treaty Association when the latter was established on 18 June 1954.

Public debates and discussions focused on NATO's activities during the Cold War, but with the dissolution of the Soviet Union (and with it the Warsaw Pact) the ATA's focus expanded. The ATA examines security issues related to Central and Eastern European countries, the Balkans, North Africa and the Middle East, as well as the Caucasus and Central Asia.

The creation of the YATA in 1996 enabled the organisation to tailor communication specifically toward younger generations in an effort to raise awareness, while continuing to work with other opinion multipliers across the Euro-Atlantic region and beyond. In 2018, the ATA launched three task forces to provide support in areas of strategic interest for the Association and the Alliance: the ATA Task Force Women, Peace and Security, dedicated to empowering women and supporting balanced gender inclusiveness in the field of defence and security ; the ATA-YATA Integrated Task Force for Communication, which uses the potential multiplier effect of the network of ATAs and YATAs in nearly 40 countries; and the ATA Task Force on Disinformation and Malign Influence, which principally analyses trends and provides training and capacity-building to ATA and YATA Chapters, as well as recommendations.

## NATO'S WIDE ACTIVITIES

### Arms control and disarmament

*I.- NATO's role in conventional arms control:*

NATO is committed to and attaches great importance to conventional arms control. The Alliance provides an essential consultative and decision-making forum for its members on all aspects of arms control and disarmament.

A) Conventional Armed Forces in Europe Treaty:

The 1990 Treaty on Conventional Armed Forces in Europe (CFE) is referred to as a "cornerstone of European security" and imposes for the first time in European history legal and verifiable limits on the force structure of its 30 States Parties, which stretch from the Atlantic Ocean to the Ural Mountains.

Since the Treaty's entry into force in 1992, the destruction of over 100,000 pieces of treaty-limited equipment (tanks, armored personnel carriers, artillery, attack helicopters and combat aircraft) has been verified and almost 6,000 on-site inspections have been conducted, thereby reaching its objective of creating balance and mitigating the possibility of surprise conventional attacks within its area of application.

At the first CFE Review Conference in 1996, negotiations began to adapt the CFE Treaty to reflect the realities of the post-Cold War era. This process was completed in conjunction with the Organization for Security and Co-operation in Europe (OSCE) Summit in Istanbul in 1999.

States Parties also agreed to additional commitments, called the Istanbul Commitments. Although the Adapted CFE (ACFE) Treaty went far in adjusting the Treaty to a new security environment, it was not ratified by Allied countries because of the failure of Russia to fully meet commitments regarding withdrawal of Russian forces from Georgia and the Republic of Moldova, on which Allies' agreement to the Adapted Treaty was based.

Since 2000 at NATO summits and ministerial meetings, the Allies have reiterated their commitment to the CFE Treaty and have reaffirmed their readiness and commitment to ratify the Adapted Treaty.

During the third CFE Review Conference in June 2006, Russia expressed its concerns regarding ratification of the adapted CFE Treaty and claimed that even the ACFE was outdated.

After the June 2007 Extraordinary Conference of the States Parties to the CFE Treaty, the Russian president signed legislation on 14 July 2007 to unilaterally "suspend" its legal obligations under the CFE Treaty as of 12 December 2007. In response to these events, NATO offered a set of constructive and forward-looking actions.

In 2008 and 2009, consultations were held between the United States (on behalf of the Alliance) and Russia, but with limited development. Further efforts to resolve the impasse were pursued on the basis of the United States' initiative, which sought an agreement on a framework for negotiations on a modernised CFE Treaty, in consultations at 36 between all CFE States Parties and NATO member states not party to the CFE Treaty. The process stalled in the autumn of 2011 because of the lack of agreement among parties.

In a situation where no agreement could be reached to overcome the impasse, towards the end of November 2011, NATO CFE Allies announced their decisions to cease implementing certain CFE obligations vis-à-vis Russia, while still continuing to fully implement their obligations with respect to all other CFE States Parties. However, in the December 2011 foreign ministers 'communiqué, Allies stated that these decisions were reversible should the Russian Federation return to full implementation.

At the Chicago Summit in May 2012, Allies reiterated their commitment to conventional arms control and expressed determination to preserve, strengthen and modernise the conventional arms control regime in Europe, based on key principles and commitments.

At the Wales Summit in September 2014, Allies reaffirmed their long-standing commitment to conventional arms control as a key element of Euro-Atlantic security and emphasised the importance of full implementation and compliance to rebuild trust and confidence. They underscored that Russia's unilateral military activity in and around Ukraine has undermined peace, security and stability across the region, and its selective implementation of the Vienna Document and Open Skies Treaty and long-standing non-implementation of the CFE Treaty have eroded the positive contributions of these arms control instruments. Allies called on Russia to fully adhere to its commitments. On 11 March 2015, the Russian Federation announced that it was suspending its participation in the meetings of the

Joint Consultative Group (JCG) on the CFE Treaty, which meets regularly in Vienna.

B) Vienna Document:

The Vienna Document (VD), that includes all European and Central Asian participating States, is a politically binding agreement designed to promote mutual trust and transparency about a state's military forces and activities. Under the VD, thousands of inspections and evaluation visits have been conducted as well as airbase visits and visits to military facilities; also new types of major weapon and equipment systems have been demonstrated to the participating States of the VD.

In 2019, NATO Allies, together with Finland and Sweden, introduced a new proposal to modernise the Vienna Document. The proposal aims to restore confidence, build mutual predictability, reduce risks and help prevent unintentional conflict.

C) Open Skies Treaty:

The Open Skies Treaty is legally binding and allows for unarmed aerial observation flights over the territory of its participants. So far, more than 1,500 observation missions have been conducted since the Treaty's entry into force in January 2002. Aerial photography and other material from observation missions provide transparency and support verification activities carried out on the ground under other treaties.

This Treaty provides for extensive cooperation regarding the use of aircraft and their sensors, thereby adding to openness and confidence. Following long-lasting negotiations the States Parties to the Open Skies Treaty agreed, at the 2010 review conference, to allow the use of digital sensors in the future. However, these have to undergo a certification process, as foreseen by the Open Skies Treaty.

D) UN Program of Action to Prevent, Combat and Eradicate the Illicit Trade in Small Arms and Light Weapons in All Its Aspects:

The proliferation of small arms and light weapons (SALW) not only feeds global terrorist activities, but also encourages violence, thus affecting local populations and preventing constructive development and economic activities.

SALW proliferation needs to be addressed as broadly as possible and the Euro-Atlantic Partnership Council (EAPC) is a well-suited framework for that. The NATO/EAPC Ad Hoc Working Group on SALW and Mine Action contributes to international efforts to address the illicit trade in SALW and encourages full implementation of international regulations and standards, including the United Nations Program of Action (UN PoA).

The UN PoA was adopted in July 2001 by nearly 150 countries, including all NATO member countries, and contains concrete recommendations for improving national legislation and controls over illicit small arms, fostering regional cooperation and promoting international assistance and cooperation on the issue. It was developed and agreed as a result of the growing realisation that most present-day conflicts are fought with illicit small arms and light weapons, and that their widespread availability has a negative impact on international peace and security, facilitates violations of international humanitarian law and human rights, and hampers economic and social development. It includes measures at the national, regional and global levels, in the areas of legislation, destruction of weapons that were confiscated, seized, or collected, as well as international cooperation and assistance to strengthen the ability of states in identifying and tracing illicit arms and light weapons. The UN holds the Biennial Meeting of States to Consider the Implementation of the PoA, in which NATO participates. National delegations from all member states gather every six years to review the progress made in the implementation of the PoA.

E) Mine action:

Although not all member states of the Alliance are a party to the Ottawa Convention on anti-personnel mines, they all fully support its humanitarian demining goals.

The Alliance assists partner countries in the destruction of surplus stocks of mines, arms and munitions through a NATO/Partnership Trust Fund mechanism.

The EAPC Ad Hoc Working Group on SALW and Mine Action also supports mine action efforts through these Trust Fund projects, as well as through information-sharing. In particular, its guest speaker program provides an opportunity for mine action experts to share their expertise with the Working Group. These speakers originate from national mine action centers, non-governmental organisations and international organisations and have included high-profile experts, such as Nobel Laureate Ms Jody Williams, Director of the International Campaign to Ban Landmines. The Group has broadened its focus to also incorporate issues related to explosive remnants of war and cluster munitions onto its agenda.

F)  Convention on Cluster Munitions:

The Convention on Cluster Munitions prohibits all use, stockpiling, production and transfer of cluster munitions. Separate articles in the Convention concerning assistance to victims, clearance of contaminated areas and destruction of stockpiles. It became a legally binding international instrument when it entered into force on 1 August 2010. As of September 2019, a total of 108 had signed the Convention while 107 States Parties had acceded to it.

G) Arms Trade Treaty:

In July 2012, A member states gathered in New York to negotiate an arms trade treaty that would establish high common standards for international trade in conventional arms.

After two years of negotiations, the Conference reached an agreement on a treaty text. Governments signed the treaty and after ratification of 50 states it came into force in December 2014. Since then, over 100 states have ratified the treaty. It establishes common international standards for the import, export and transfer of conventional arms. NATO stands ready to support the Arms Trade Treaty as necessary.

H) Trust Fund projects:

The Partnership for Peace Trust Fund mechanism was originally established in 2000 to assist partner countries with the safe destruction of stocks of anti-personnel land mines.

It was later extended to include the destruction of surplus munitions, unexploded ordnance and SALW, and assisting partner countries in managing the consequences of defence reform, training and building integrity. So far, NATO has contributed to the destruction of 5.65 million anti-personnel landmines, 46,750 tonnes of various munitions, 2 million hand grenades, 15.95 million cluster submunitions, 1,635 man-portable air defence systems (MANPADS), 3,530 tonnes of chemicals and 626,000 SALW, alongside 164.4 million rounds of SALW ammunition.

Over the years, NATO has trained thousands of explosive ordnance disposal experts, giving, for instance, assistance to more than 12,000 former military personnel through defence reform Trust Fund projects.

Trust Fund projects are initiated by a NATO member or partner country and funded by voluntary contributions from individual Allies, partners and organisations. A web-based information-sharing platform allows donors and recipient countries to share information about ongoing and potential projects.

NATO bodies involved in conventional arms control:

There are a number of NATO bodies that provide a forum to discuss and take forward arms control issues. Arms control policy is determined within the deliberations of the High-Level Task Force (HLTF) on Conventional Arms Control that was established for CFE and confidence- and security-building measures (CSBMs).

Implementation and verification of arms control agreements fall under the purview of the Verification Coordinating

Committee (VCC), including overseeing a designated CFE verification database.

Other fora include the Partnerships and Cooperative Security Committee (PCSC) and the EAPC Ad Hoc Working Group on SALW and Mine Action, in which implementing organisations like the UN, the European Union, the OSCE, the South Eastern and Eastern Europe Clearinghouse for the Control of SALW (or SEESAC) and the NATO Support and Procurement Agency (NSPA) can share information on projects.

The NATO-Russia Council (NRC) also has a working group for Arms Control, Disarmament and Non-Proliferation. However, work of the NRC has been suspended since spring 2014 due to Russia's actions in Ukraine.

The NATO School in Oberammergau (Germany) conducts several courses in the fields of arms control, disarmament and non-proliferation. They are related to CFE, VD, Open Skies, weapons of mass destruction (WMD), SALW and Mine Action. Most of them are also open to NATO's partners.

*II.-Arms control, disarmament and non-proliferation in NATO:*

NATO has a long-standing commitment to an active policy in arms control, disarmament and non-proliferation. The Alliance continues to pursue its security objectives through this policy, while at the same time ensuring that its collective defence obligations are met and the full range of its missions fulfilled.

A) Definitions:

While often used together, the terms arms control, disarmament and non-proliferation do not mean the same thing. In fact, experts usually consider them to reflect associated, but different areas in the same discipline or subject.

a) Arms control

Arms control is the broadest of the three terms and generally refers to mutually agreed upon restraints or controls (usually between states) on the development, production, stockpiling, proliferation, deployment and use of troops, small arms, conventional weapons and weapons of mass destruction.

Arms control includes agreements that increase the transparency of military capabilities and activities, with the intention of reducing the risk of misinterpretation or miscalculation.

b) Disarmament

Disarmament, often inaccurately used as a synonym for arms control, refers to the act of eliminating or abolishing weapons (particularly offensive arms) either unilaterally (in the hope that one's example will be followed) or reciprocally.

It may refer either to reducing the number of arms, or to eliminating entire categories of weapons.

c) Non-proliferation

For the Alliance, "non-proliferation refers to all efforts to prevent proliferation from occurring, or should it occur, to reverse it by any other means than the use of military force." Non-proliferation applies to both weapons of mass destruction (including nuclear, chemical and biological weapons) and conventional capabilities such as missiles and small arms.

d) Weapons of mass destruction proliferation

Attempts made by state or non-state actors to develop, acquire, manufacture, possess, transport, transfer or use nuclear, chemical or biological weapons or devices and their means of delivery or related material, including precursors, without prejudice to the rights and obligations of the States Parties to the following agreements: the Treaty on the Non-Proliferation of Nuclear Weapons or Non-Proliferation Treaty (NPT), the Convention on the Prohibition of the Development, Production, Stockpiling and Use of Chemical Weapons and on their Destruction (CWC) and the Convention on the Prohibition of the Development, Production and Stockpiling of Bacteriological (Biological) and Toxin Weapons and on their Destruction (BTWC).

B) How NATO contributes:

"We need to preserve and implement the Non-Proliferation Treaty. We need to adapt nuclear arms control regimes to new realities. We need to modernise the Vienna Document.

And we need to consider how to develop new rules and standards for emerging technologies, including advanced missile technology". These are the four areas of activity outlined by NATO Secretary General Jens Stoltenberg, in October 2019, where Allies will act together in support of arms control, disarmament and non-proliferation. NATO contributes in many ways to these efforts through its policies, its activities and through its member countries.

NATO's policies in these fields cover consultation and practical cooperation in a wide range of areas. They include conventional arms control; nuclear policy issues; promoting mine action and combatting the spread of small arms and light weapons (SALW); preventing the proliferation of weapons of mass destruction (WMD); and developing and harmonising capabilities to defend against chemical, biological, radiological and nuclear (CBRN) threats.

1) Conventional forces:

Allies have reduced their conventional forces significantly from Cold War levels. NATO Allies that are Parties to the CFE Treaty remain committed to the regime of the Treaty. As a response to Russia`s unilateral "suspension" of its Treaty obligations in 2007, NATO CFE Allies ceased implementing certain Treaty obligations vis-à-vis Russia in November 2011, while still continuing to implement fully their obligations with respect to all other CFE States Parties. Allies stated that these decisions are fully reversible should Russia return to full implementation. At their summits since 2014, Allies have reaffirmed their long-standing commitment to conventional arms control as a key element of Euro-Atlantic security and emphasised the importance of full implementation and compliance to rebuild trust and confidence. They underscored that Russia's unilateral military activity in and around Ukraine has undermined peace, security and stability across the region, and its selective implementation of the Vienna Document and Open Skies Treaty and long-standing non-implementation of the CFE Treaty have eroded the positive contributions of these arms control instruments. Allies called on Russia to fully adhere to its commitments. In October 2019, NATO Allies submitted the most comprehensive modernisation proposals of the Vienna Document since 1994. These proposals have been presented to all participating states of the OSCE in Vienna.

2) Nuclear forces:

NATO is committed to the goal of creating the conditions for a world without nuclear weapons, but reconfirms that, as long as there are nuclear weapons in the world, NATO will remain a nuclear alliance. However, it will do so at the lowest level consistent with its defence obligations, and with an appropriate mix of nuclear and conventional forces. The nuclear weapons committed to NATO have been reduced by more than 90 per cent since the height of the Cold War, and the role of nuclear weapons in NATO's defence doctrine has been dramatically reduced.

NATO nuclear weapon states have also reduced their nuclear arsenals and ceased production of highly enriched uranium or plutonium for nuclear weapons. Allies remain committed to crafting the conditions for further reductions in the future on the basis of reciprocity, recognising that progress on arms control and disarmament must take into account the prevailing international security environment.

Allies also emphasise their strong commitment to full implementation of the Nuclear Non-Proliferation Treaty (NPT) in all its aspects, including nuclear disarmament, non-proliferation and the peaceful uses of nuclear energy, science and technology. The NPT has been the cornerstone of global non-proliferation and disarmament efforts for 50 years, and has an essential role in the maintenance of international peace, security and stability. On the occasion of the 50th anniversary of the Treaty on 5 March 2020, NATO issued a statement in which Allies reaffirmed their "resolve to seek a safer world for all, and to take further practical steps and effective measures to foster nuclear disarmament."

Allies stress, in this statement, the unique role played by the treaty and in view of ongoing proliferation issues "call on all States to enhance efforts to achieve universal adherence and universalisation, and effectively combat nuclear proliferation through full implementation of the NPT". NATO Allies continue by stating that they "support the ultimate goal of a world without nuclear weapons in full accordance with all provisions of the NPT, including Article VI, in an ever more effective and verifiable way that promotes international stability, and is based on the principle of undiminished security for all".

NATO Allies were also strongly in favor of preserving the Intermediate-Range Nuclear Forces (INF) Treaty. Despite years of US and Allied engagement, Russia continued to develop and deploy the SSC-8/9M729 missile system, which violated the Treaty. In December 2018, NATO Foreign Ministers supported the finding of the United States that Russia was in material breach of its obligations under the INF Treaty and called on Russia to urgently return to full and verifiable compliance. Russia, nevertheless, continued to deny its Treaty violation. As a consequence, on 1 February 2019, the United States suspended its obligations under the INF Treaty. The American withdrawal from the Treaty took effect on 2 August 2019, six months after this announcement. During this six month period, Russia continued to deny its Treaty violation and did not honor its obligations through the verifiable destruction of its SSC-8/9M729 system. As such, Allies agree that Russia bears sole responsibility for the demise of the Treaty.

NATO has agreed to balanced, coordinated and defensive package of measures to ensure that the Alliance's deterrence and defence remains credible and effective in the face of the significant risks posed by Russia's missile system. Allies remain firmly committed to the preservation of effective international arms control, disarmament and non-proliferation.

3)  Armed forces:

Through its cooperation framework with non-member countries, the Alliance supports defence and security sector reform, emphasising civilian control of the military, accountability, and restructuring of military forces to lower, affordable and usable levels.

4) Small arms and light weapons (SALW) and mine action (MA):

Allies are working with non-member countries and other international organisations to support the full implementation of the UN Program of Action to Prevent, Combat and Eradicate the Illicit Trade in SALW in All its Aspects.

NATO also supports mine action activities across the globe. All NATO member countries, with the exception of the United States, are party to the 1997 Mine Ban Treaty, often referred to as the Ottawa Convention.

NATO's Partnership for Peace (PfP) Trust Fund Policy was initiated in 2000 to assist countries in fulfilling their Ottawa Convention obligations to dispose of stockpiles of anti-personnel landmines. The policy was later expanded to include efforts to implement the UN Program of Action on SALW. More recently, the Trust Fund Policy has also been expanded to include projects addressing the consequences of defence reform, training and building integrity.

NATO/Partnership Trust Funds may be initiated by a NATO member or partner country to tackle specific, practical issues linked to these areas. They are funded by voluntary contributions from individual NATO Allies, partners and organisations.

At the 2018 Brussels Summit, NATO Heads of State and Government emphasised the need to do more to achieve lasting calm and an end to violence in the Middle East and North Africa, which face continuing crises and instability with direct implications for the security of NATO. They also made a plea for enhanced practical cooperation, including through further support in the areas of counter-terrorism, small arms and light weapons, countering improvised explosive devices, and military border security.

5) Weapons of mass destruction (WMD):

"With due respect to the primarily military mission of the Alliance, NATO will work actively to prevent the proliferation of WMD by State and non-State actors, to protect the Alliance from WMD threats should prevention fail, and be prepared for recovery efforts should the Alliance suffer a WMD attack or CBRN event, within its competencies and whenever it can bring added value, through a comprehensive political, military and civilian approach".

NATO stepped up its activities in this area in 1999 with the launch of the WMD Initiative and the establishment of a WMD Center at NATO Headquarters the following year. NATO Allies engage in preventing the proliferation of WMD by state and non-state actors through an active political agenda of arms control, disarmament and non-proliferation. They also do this by developing and harmonising defence capabilities and, when necessary, by employing these capabilities, consistent with political decisions in support of non-proliferation objectives. Both political and defence elements are essential to NATO's security, as well as the preparedness for recovery efforts, should it suffer a WMD attack or CBRN event.

The Alliance engages actively to enhance international security through partnership with relevant countries and other international organisations. NATO's partnership programs are therefore designed to provide effective frameworks for dialogue, consultation and coordination. They actively contribute to NATO's arms control, non-proliferation and disarmament efforts.

Of particular importance is the outreach to and cooperation with the United Nations (UN), the European Union (EU), and other organisations and multilateral initiatives that address WMD proliferation.

Since 2004, NATO organises the annual non-proliferation conference. This unique event provides a venue for senior national officials to informally discuss WMD threats. The last WMD Conference was held in Brussels, Belgium in October 2019.

6) Chemical weapons:

Since its entry into force in 1997, the Chemical Weapons Convention has become one of the pillars of the global non-proliferation regime. The Convention prohibits the development, transfer and use of chemical weapons. States Parties to the Convention include all NATO member countries; they commit not to develop, produce or acquire, stockpile or retain chemical weapons, nor to transfer, directly or indirectly, chemical weapons to anyone. States Parties also undertake not to engage in any military preparations to use chemical weapons, nor to commit to assist, encourage or induce anyone to engage in prohibited activity.

The first offensive use of a nerve agent on Alliance territory since NATO's foundation occurred on 4 March 2018 in Salisbury, the United Kingdom. The military grade nerve agent was of a type developed by Russia. Allies agree that the attack was a clear breach of international norms and agreements, and they have called on Russia to disclose the Novichok program to the Organization for the Prohibition of Chemical Weapons.

Allies strongly condemned the repeated use of chemical weapons by the Syrian regime and called for those responsible to be held to account. Despite sustained diplomatic efforts, the Syrian regime's repeated use of chemical weapons against civilians contributed to appalling human suffering since the start of the conflict in 2011. The use of such weapons was in flagrant violation of international standards and non-proliferation norms, multiple UN Security Council Resolutions, and the Chemical Weapons Convention, which Syria ratified in 2013. NATO considers any use of chemical weapons by state or non-state actors to be a threat to international peace and security.

Overview of NATO's contribution:

Active policies in arms control, disarmament and non-proliferation have been an inseparable part of NATO's contribution to security and stability since 1957 when Allies put forward the first NATO Disarmament Proposal in London and subsequently established regular meetings of disarmament experts at NATO Headquarters. These policies were most clearly articulated in the Harmel Report of 1967.

a) Harmel Report:

- This report forms the basis for NATO's security policy. It outlined two objectives: maintaining a sufficient military capacity to act as an effective and credible deterrent against aggression and other forms of pressure, and, on that basis, seeking to improve East-West relations through dialogue. The Alliance's objectives in arms control are tied to the achievement of both aims. Deterrence and defence help to prevent war against hostile adversaries; arms control and dialogue help to prevent war when the sides find a common interest in avoiding conflict. It is therefore important that deterrence and defence policies, and arms control policies remain mutually reinforcing with the common goal of preventing war.

b) Comprehensive Concept of Arms Control and Disarmament:

- In May 1989, NATO adopted the Comprehensive Concept of Arms Control and Disarmament, which allowed the Alliance to move forward in the sphere of arms control. It addressed the role of arms control in East-West relations, the principles of Alliance security and a number of guiding principles and objectives governing Allied policy in the nuclear, conventional and chemical fields of arms control.

- It clearly set out the interrelationships between arms control and defence policies and established the overall conceptual framework within which the Alliance sought progress in each area of its arms control agenda.

c) The Alliance's Strategic Concept:

- NATO's continued adherence to this policy was reaffirmed in the 2010 Strategic Concept (with regard to nuclear weapons):

- "It [This Strategic Concept] commits NATO to the goal of creating the conditions for a world without nuclear weapons; but reconfirms that, as long as there are nuclear weapons in the world, NATO will remain a nuclear Alliance".

- "NATO seeks its security at the lowest possible level of forces. Arms control, disarmament and non-proliferation contribute to peace, security and stability, and should ensure undiminished security for all Alliance members. We will continue to play our part in reinforcing arms control and in promoting disarmament of both conventional weapons and weapons of mass destruction, as well as non-proliferation efforts".

d) Deterrence and Defence Posture Review:

- The NATO Deterrence and Defence Posture Review (DDPR), agreed at the Chicago Summit in 2012, addresses issues of arms control, disarmament and non-proliferation. The DDPR document underscores: "The Alliance is resolved to seek a safer world for all and to create the conditions for a world without nuclear weapons in accordance with the goals of the Nuclear Non-Proliferation Treaty, in a way that promotes international stability, and is based on the principle of undiminished security for all". It also repeats that as long as nuclear weapons exist, NATO will remain a nuclear alliance.

- The Special Advisory and Consultative Arms Control, Disarmament and Non-Proliferation Committee (ADNC) was established on the basis of DDPR agreement.

e) Summit declarations:

- Allied leaders have reiterated their commitment to arms control, disarmament, and non-proliferation since the first NATO Summit in 1957.

- At the meeting of NATO leaders in London in 2019, Allies reiterated their full commitment to the preservation and strengthening of effective arms control, disarmament and non-proliferation, taking into account the prevailing security environment. Allies also expressed their strong commitment to the full implementation of the NPT in all its aspects, including nuclear disarmament, non-proliferation, and the peaceful uses of nuclear energy.

- At the Brussels Summit in 2018, Allies reiterated their long-standing positon that arms control, disarmament and non-proliferation have made, and should continue to make, an essential contribution to achieving the Alliance's security objectives and for ensuring strategic stability and our collective security. NATO has a long track record of doing its part on disarmament and non-proliferation. Allies expressed their position that the Intermediate-Range Nuclear Forces (INF) Treaty has been crucial to Euro-Atlantic security. Furthermore, the Allies also underlined the importance of effective multilateralism and international cooperation, including through the Chemical Weapons Convention and the Organization for the Prohibition of Chemical Weapons (OPCW), in addressing WMD threats. In that spirit, NATO welcomed the decision by the June 2018 OPCW Conference of States Parties, in particular to ask the independent experts of the OPCW Technical Secretariat to put in place arrangements to identify the perpetrators of the use of chemical weapons in Syria. Allies demanded that all perpetrators of chemical weapons attacks worldwide be held accountable and called upon all countries to join the International Partnership against Impunity for the Use of Chemical Weapons.

- In the 2016 Warsaw Summit Declaration, the Alliance reaffirmed its long-standing commitment to conventional arms control as a key element of Euro-Atlantic security and emphasised the importance of full implementation and compliance to rebuild trust and confidence. Allied leaders also stated that Russia's unilateral military activity in and around Ukraine has undermined peace, security and stability across the region, and its selective implementation of the Vienna Document and

Open Skies Treaty and long-standing non-implementation of the Conventional Armed Forces in Europe (CFE) Treaty have eroded the positive contributions of these arms control instruments. At Warsaw, NATO also continued to call on Russia to preserve the viability of the INF Treaty and condemned the Democratic People's Republic of Korea (DPRK) for its multiple ballistic missile tests and its nuclear tests, calling DPRK to immediately cease and abandon all its existing nuclear and ballistic missile activities in a complete, verifiable, and irreversible manner and re-engage in international talks.

- At the 2009 Strasbourg/Kehl Summit, Allied leaders endorsed NATO's Comprehensive, Strategic-Level Policy for Preventing the Proliferation of Weapons of Mass Destruction and Defending Against Chemical, Biological, Radiological and Nuclear Threats. Additionally, in 2008 at the Bucharest Summit, Allied leaders took note of a report on raising NATO's profile in the fields of arms control, disarmament and non-proliferation. As part of a broader response to security issues, they agreed that NATO should continue to contribute to international efforts in these fields and keep these issues under active review, something they subsequently did at each one of the ensuing summits.

NATO committees on arms control disarmament, and non-proliferation:

A number of NATO committees and bodies oversee different aspects of Alliance activities in the fields of arms control, disarmament and non-proliferation. Within the International Staff, the Arms Control, Disarmament and WMD Non-Proliferation Center (ACDC) oversees the committees that deal with arms control, disarmament and non-proliferation. As with all policy, overall political guidance is provided by the North Atlantic Council, NATO's highest political decision-making body. Overall military guidance is provided by the Military Committee.

Detailed oversight of activities and policy in specific areas is provided by a number of bodies, including the High Level Task Force (HLTF) on Conventional Arms Control; the Special Advisory and Consultative Arms Control, Disarmament and Non-Proliferation Committee (ADNC); the Nuclear Planning Group High Level Group (NPG/HLG); the Verification Coordinating Committee (VCC); and the Committee on Proliferation (CP) in Politico-Military format (for global non- and counter-proliferation issues) and Defence format (for CBRN defence issues). The Euro-Atlantic Partnership Council and the Political Committee also both meet on small arms and light weapons and mine action topics in ad hoc formats, coordinating Allied and partner work on mitigating this important global threat.

*III.- Small arms and light weapons (SALW) and mine action (MA):*

The proliferation of small arms and light weapons (SALW) affects security while anti-personnel mines and explosive remnants of war kill and maim both people and livestock long after the end of hostilities. Both can have destabilising effects on social, societal and economic development and can represent major challenges to regional and national security.

The challenges posed by SALW and mines:

The illicit proliferation of SALW can fuel and prolong armed violence and support illegal activities and the emergence of violent groups. Access to illicit SALW contributes to the development of terrorism, organised crime, human trafficking, gender violence and piracy; and the diversion of weapons is closely linked to corruption and poor management practices. Small arms are weapons intended for use by an individual. They include pistols, rifles, submachine guns, assault rifles and light machine guns; light weapons are designed for use by two or more persons serving as a crew and include heavy machine guns, grenade launchers, mortars, anti-aircraft guns and anti-tank guns, all less than 100 mm in caliber.

Anti-personnel mines and explosive remnants of war kill and maim both people and livestock long after the cessation of hostilities and are a major barrier to post-conflict recovery and development. Beyond the human tragedy they can cause, they also overload local and national health services, reduce the available workforce and disrupt the social and societal structures. In many countries, stockpiles of weapons and ammunition are not always properly managed, allowing illicit access or accidents that may affect security personnel and nearby populations.

NATO is helping to address these issues by encouraging dialogue and cooperation among Allies and partners to seek effective solutions. It has two very effective mechanisms:

the Ad Hoc Working Group on SALW and Mine Action (AHWG SALW/MA) and the NATO/Partnership Trust Fund mechanism. NATO also supports initiatives led by other international bodies, such as the United Nations (UN) Program of Action to Prevent, Combat, and Eradicate the Illicit Trade in SALW in All Its Aspects (commonly known as the PoA) as well as the UN Arms Trade Treaty (ATT).

In the area of anti-personnel mines, the Alliance and its partners also assist signatories of the "Convention on the Prohibition of the Use, Stockpiling, Production and Transfer of Anti-Personnel Mines and Their Destruction" (Ottawa Convention). Allies who are not party to this Convention facilitate efforts in the general realm of what is commonly called mine action, which includes: clearance of mine fields, providing victim assistance, raising mine risk awareness through education, and assistance in destroying mine stockpiles.

Tackling both issues together:

In 1999, the Euro-Atlantic Partnership Council (EAPC), which groups Allies and partner countries, established the AHWG on SALW. Originally, this Working Group focused only on issues related to the impact of the proliferation of SALW on Alliance's peacekeeping operations.

In April 2004, the Working Group's mandate was broadened to include mine action issues (therefore becoming the AHWG SALW/MA). It is one of the few forums in the world that meets on a regular basis to address these specific issues. The objective of the Working Group is to contribute to international efforts to reduce the threats caused by the illicit trade of SALW and the impact of mines and other unexploded ordnance.

a) An annual work program:

- The Working Group organises itself around an annual work program. In practice, it uses a four-pronged approach to accomplish its work by:

  1. providing a forum in which EAPC members and certain implementing organisations can share information on SALW and ammunition projects they are conducting. These organisations include but are not limited to the European Union (EU), the NATO Support and Procurement Agency (NSPA), the Organization for Security and Co-operation in Europe (OSCE), the South Eastern and Eastern Europe Clearinghouse for the Control of Small Arms and Light Weapons (SEESAC) and the United Nations (UN). This exchange of information helps to improve coordination with donor countries and implementing organisations, with the aim of increasing effectiveness and avoiding duplication of work. The information is consolidated into the Project Information Matrix, a web-based information-sharing platform, which is regularly updated by the members of the AHWG SALW/MA;

  2. inviting partners from the Mediterranean Dialogue (MD), the Istanbul Cooperation Initiative (ICI), as well as partners across the

globe, to share information and identify national and regional approaches;

3. inviting speakers from non-governmental organisations (NGOs), regional and international organisations, and research institutes to share their views and recent research with delegations;

4. facilitating the management and creation of the Trust Fund projects. This includes updating delegations on the status of Trust Fund projects and highlighting where more effort or volunteer donations are needed;

5. organising regular international workshops, seminars and conferences on topics particularly pertinent to SALW and mine action.

- NATO's International Staff (IS) functions as the Working Group's executive agent and implements the annual work programs of the AHWG SALW/MA and organisations its quarterly meetings.

b) Training:

- NATO conducts an annual course related to SALW and mine action that is usually held at the NATO School in Oberammergau, Germany. The "SALW and Mine Action Course", aimed at mid-level management personnel, provides students with an overview of the most significant political, practical and regulatory issues needed to deal with SALW and their ammunition and their life-cycle management, as well as mine action from a national, regional and global perspective. It includes cross-cutting issues, such as capacity-building, defence education, gender mainstreaming and good governance that affect the various facets of issues related to the risks and challenges caused by SALW and mine action. The course is open to military and civilian personnel from all NATO partner countries, as well as to relevant international organisations.

c) NATO support to global efforts:

- The UN Program of Action to Prevent, Combat and Eradicate the Illicit Trade in Small Arms and Light Weapons in All its Aspects (known as the PoA) was adopted in July 2001 by nearly 150 countries, including all NATO member countries. It consists of measures at the national, regional and global levels, in the areas of legislation, destruction of weapons that were confiscated, seized or collected, as well as international cooperation and assistance to strengthen the ability of states in identifying and tracing illicit arms and light weapons. The NATO Ad Hoc Working Group supports the implementation of the PoA and related international instruments through its activities and will continue to support major global events of this nature.

- On 1 August 2010, the Convention on Cluster Munitions (CCM) entered into force and became a legally binding instrument. The CCM prohibits, for its signatories, all use, stockpiling, production and transfer of cluster munitions. Separate articles in the

Convention concerning assistance to victims, clearance of contaminated areas and the destruction of stockpiles. The NATO Working Group provides an additional forum for the discussion and facilitation of its implementation.

• The landmark Arms Trade Treaty (ATT), regulating the international trade in conventional arms, from small arms to battle tanks, combat aircraft and warships, entered into force on 24 December 2014. The treaty aims to foster peace and security by interrupting the destabilising flow of arms to conflict regions. NATO supports the implementation of the ATT in particular through the activities of the Working Group on SALW and Mine Action and constitutes an additional forum for discussion and information-sharing on the issue.

Trust Funds projects:

The end of the Cold War left a dangerous legacy of aging arms, ammunition, anti-personnel mines, missiles, rocket fuel, chemicals and unexploded ordnance. In 1999, NATO established the NATO Partnership for Peace (PfP) Trust Fund mechanism to assist partners with these problems. The NATO PfP Trust Fund Policy was established in September 2000 in order to assist partners in meeting the Ottawa Convention obligations. The policy expanded to include disposal of conventional ammunition, small arms, defence reform, training and building integrity. Since then, Trust Fund projects have produced tangible results and, as such, represent the operational dimension of the Working Group's efforts.

Trust Fund projects focus on the destruction of SALW, ammunition and mines, improving their physical security and stockpile management, and also address the consequences of defence reform. Allies and partners fund and execute these projects through NSPA as the main executive agent. Each project has a lead nation (s), which oversees the development of project proposals along with the NATO International Staff and the executive agent. This ensures a mechanism with a competitive bidding process, transparency in how funds are expended and verifiable project oversight, particularly for projects involving the destruction of munitions.

Trust Funds may be initiated by a NATO member or partner country to tackle specific, practical issues linked to the demilitarisation process of a country or to the introduction of defence reform projects. They are funded by voluntary contributions from individual NATO Allies, partner countries, and more recently NGOs. They are often implemented in cooperation with other international organisations and NGOs.

As an example, until the first two decades of the 21st century, Allies and partners, through the Trust Fund projects, have destroyed or cleared:

164 million rounds of ammunition
15.9 million cluster sub-munitions
5.6 million anti-personnel landmines
2 million hand grenades
626,000 small arms and light weapons (SALW)
643,000 pieces of unexploded ordnance (UXO)
46,750 tonnes of various ammunition
94,500 surface-to-air missiles and rockets

1,635 man-portable air defence systems (MANPADS)
3,530 tonnes of chemicals, including rocket fuel oxidiser ("mélange")
4,120 hectares cleared

In addition, over 12,000 former military personnel have received retraining assistance through defence reform Trust Fund projects.

The Trust Fund mechanism is open to countries where NATO is leading a crisis management operation and to countries participating in NATO's PfP program, the Mediterranean Dialogue, the Istanbul Cooperation Initiative, and to partners from across the globe. For instance, in 2018, NATO launched the implementation of the fourth phase of the Jordan Trust Fund project on ammunition stockpile management and ammunition destruction, and continued to cooperate on physical security and stockpile management, and ammunition demilitarisation in Serbia, thus enhancing safety and security of local communities.

Once the project proposal is agreed by the lead nation and the partner country concerned, it is presented to the Partnerships and Cooperative Security Committee (PCSC), which is the formal forum to discuss projects and attract volunteer donor support and resources. The Luxembourg-based NSPA has been selected by lead nations of most Trust Fund projects to be the executing agent, particularly for demilitarisation projects. It plays a key role in the development and implementation of Trust Fund projects and offers technical advice and a range of management services.

## Countering human trafficking during military operations

The Alliance initiated a zero-tolerance policy on human trafficking, which was endorsed at the Istanbul Summit in June 2004. The policy commits NATO member countries and other troop-contributing nations participating in NATO-led operations to reinforce efforts to prevent and combat such activity. The issue is kept under regular review by the Euro-Atlantic Partnership Council (EAPC). The policy was also opened to the Mediterranean Dialogue and Istanbul Cooperation Initiative countries, as well as four partners across the globe (Australia, Japan, the Republic of Korea, the New Zealand) and remaining operational partners (Colombia, Malaysia, Mongolia, Singapore, Tonga) in January 2011.

NATO member countries are all signatories to the UN Protocol on Trafficking in Persons. The Allies are keenly aware that human trafficking fuels corruption and organised crime, and therefore runs counter to NATO's stabilisation efforts in its theaters of operation. These considerations led to the development of the NATO policy on combating trafficking in human beings.

NATO does not see itself as the primary organisation to combat trafficking in human beings, but is working to add value wherever it can. The policy was developed in consultation with EAPC countries and non-NATO troop contributors, as well as governmental and non-governmental organisations.

The zero-tolerance policy calls for military and civilian personnel and contractors taking part in NATO-led operations to receive appropriate training on standards of their behavior during the operations. The Allies also agreed to review national legislation and report on national efforts in this regard. In theater, NATO-led forces, operating within the limits of their mandate, support the responsible host-country authorities in their efforts to combat trafficking in human beings.

Much of the responsibility for implementing the policy was assigned to NATO's Military Committee given that it is troops from NATO and non-NATO nations participating in NATO-led operations who are the most likely to come into contact with trafficked individuals and trafficking rings. Guidance was then issued by the Strategic Commanders.

The policy is kept under review to make sure that it's effectively implemented by Allies and Partners as well as NATO as an organisation. A regular comprehensive review is conducted to provide policy and practical recommendations. These include measures to strengthen policies and provisions in specific operations, to enhance training and awareness raising among NATO forces as well as the evaluation and reporting of all related activities.

A Senior Coordinator on Combating Trafficking in Human Beings (the NATO ASG for Defence Policy and Planning) coordinates all Alliance efforts in this field.

Developing policies and provisions in specific operations:

The Alliance is working to ensure that the entire chain of command in every operation is aware of the NATO policy. Within existing operations the Allies are developing specific policy provisions, which do not exceed NATO's mandate, for the role of NATO-led forces in supporting the authorities of the host country in combating the trafficking of human beings.

Specific policy provisions have been developed and incorporated into the operational plans relating to Afghanistan and Balkans to reflect the NATO policy and relevant guidance, as well as to raise the awareness of personnel. The appropriate role for NATO forces in this area is to support activities to the local authorities and relevant international organisations. Maintaining close contact with the host country is vital.

In Afghanistan, the International Security Assistance Force (ISAF) is tasked to provide support to the Government of Afghanistan in countering human trafficking. ISAF works alongside and shares information with the Afghan security forces. ISAF holds weekly meetings with the International Organization for Migration, which has been designated as the lead agency on the issue by the UN Assistance Mission in Afghanistan (UNAMA). ISAF also liaises regularly with the German police project, the UNAMA Human Rights Unit, the UNAMA Gender Advisor, the UN High Commissioner for Refugees (UNHCR), the United Nations Children's Fund (UNICEF) and the Afghanistan Independent Human Rights Commission.

In Kosovo, the United Nations Mission in Kosovo (UNMIK) has the lead on the issue. The Kosovo Force (KFOR) supports the UNMIK police (UNMIK-P) which has the executive responsibility.

Training and raising awareness among Allies forces:

Training and raising awareness among NATO forces is essentially the responsibility of the individual troop-contributing nation. Yet the Alliance is addressing the issue in a number of courses for the military personnel of both NATO and Partner countries at the NATO Defence College in Rome and NATO School in Oberammergau (NSO), Germany. Options for enhancing training in this area are being considered. The NSO also provides two Advanced Distant Learning courses related to combating trafficking in human beings, which are available to all those that may want to use them. Moreover, since 2008, the Turkish PfP Training Center organises a bi-annual course on "Fight Against Trafficking in Human Beings", which is open to military and civilians from NATO, PfP, MD and ICI countries.

Accountability under the zero-tolerance policy:

Nations contributing troops to NATO-led operations are required to ensure that members of their forces (as well as civilian elements) who engage in human trafficking or facilitate it, are liable to appropriate prosecution and punishment under their national legislation. Senior NATO commanders could ask for the repatriation of any offenders.


**Gender issues**

*I.- Women, Peace and Security:*

NATO demonstrates its commitment to gender equality through the implementation of United Nations Security Council Resolutions (UNSCRs) on Women, Peace and Security (WPS). These Resolutions (1325, 1820, 1888,

1889, 1960, 2106, 2122 and 2422) recognise the disproportionate impact that conflict has on women and girls, and call for full and equal participation of women at all levels of conflict prevention to post-conflict reconstruction, and protection of women and girls from sexual violence in conflict.

Responding to the call for action:

The WPS mandate is fundamental to NATO's common values of individual liberty, democracy, human rights and obligations under the Charter of the United Nations. In line with the UNSCRs on WPS, NATO aims to address gender inequality and integrate WPS priorities through the Alliance's three core tasks of collective defence, crisis management and cooperative security.

NATO is actively seeking to incorporate gender perspectives within the analysis, planning, execution and evaluation of its operations and missions. This is also an important focus in NATO's cooperation with partner countries, both in the preparation of troops that will deploy in NATO-led operations and missions, as well as in wider cooperation on defence capacity building. NATO is also seeking to promote greater gender equality and increase the participation of women in defence and security institutions within the Organization and its member countries.

NATO cooperates with other international organisations to advance the overall agenda on WPS. The Regional Acceleration of Resolution 1325 (RAR) framework serves as a joint platform for NATO, the EU, OSCE, UN and AU for sharing best practices on WPS. NATO also recognises the important role civil society organisations continue to play in overseeing the promotion of women's and girls' empowerment and the protection of their rights. To better support NATO's implementation of the UNSCRs on WPS, the Civil Society Advisory Panel (CSAP) was established. The CSAP provides overarching recommendations on the integration of a gender perspective into NATO's core tasks and liaises with women's organisations in national settings.

A number of gender-related projects under the NATO Science for Peace and Security (SPS) Program involve civil networks of experts from Allied and partner countries, providing a forum for sharing knowledge and solving issues of common interest.

Overarching policy and action plan:

NATO and its partners' active commitment to the UNSCRs on WPS resulted in a formal NATO/EAPC Policy on Women, Peace and Security to support the implementation of these Resolutions, first issued in December 2007.

A first Action Plan to support the implementation of this Policy was endorsed at the Lisbon Summit in 2010 on the occasion of the 10th anniversary of UNSCR 1325.

The Action Plan has been revised on a biannual basis since 2014 to reflect its implementation. The Policy and the Action Plan were both revised in 2018, ahead of endorsement by Heads of State and Government at the Brussels Summit in July 2018. The Allies, together with their EAPC partners, as well as Afghanistan, Australia, Japan, Jordan, New Zealand and the United Arab Emirates have signed up for their implementation. Other interested partners will be invited to also adhere to the revised Policy and Action Plan following the Brussels Summit.

In the NATO/EAPC Policy on WPS, NATO and its partners recognise the adoption of the WPS agenda and support the advancement of gender equality through the guiding principles of:

Integration: gender equality must be considered as an integral part of NATO policies, programs and projects guided by effective gender mainstreaming practices.

To achieve gender equality, it must be acknowledged that each policy, program, and project affects both women and men.

Inclusiveness: representation of women across NATO and in national forces is necessary to enhance operational effectiveness and success. NATO will seek to increase the participation of women in all tasks throughout the International Military Staff and International Staff at all levels.

Integrity: systemic inequalities are addressed to ensure fair and equal treatment of women and men Alliance-wide. Accountability on all efforts to increase awareness and implementation of the WPS agenda will be made a priority in accordance with international frameworks.

NATO and its partners aim to contribute to the implementation of the UNSCRs on WPS by making this Policy an integral part of their everyday business in both civilian and military structures.

a)  Working with partner countries:

- Through their cooperation programs with NATO, partners are encouraged to adopt specific goals that reflect the principles and support implementation of the UNSCRs on WPS.

- They are also invited to make use of the training and education activities developed by Allied Command Transformation, which has ensured that a gender perspective is included in the curriculum of NATO Training Centers and Centers of Excellence as well as in pre-deployment training.

- Though the Alliance has no influence on measures or policies taken at national levels, all personnel - whether from Allied or partner countries - deployed in NATO-led operations and missions or serving within NATO structures must be appropriately trained and meet required standards of behavior. Several countries have initiated gender-related training for subject matter experts and raised general awareness on the UNSCRs on WPS ahead of national force deployments.

- Work among Allies and partner countries is not only about developing gender awareness in crisis-management or peace-support operations. An increasingly important focus is on strengthening gender perspectives, including promoting gender equality and the participation of women in defence and security institutions, as well as in the armed forces.

b) Gender perspective in operations:

- WPS Resolutions are also being implemented in crisis management and in NATO-led operations and missions. The Alliance has nominated gender advisers at both Strategic Commands - Allied Command Operations and Allied Command Transformation, as well as in subordinate commands and in NATO-led operations and missions. Gender advisers support commanders to ensure that gender perspectives are integrated in all aspects of an operation.

- In 2015, NATO and its partners adopted the Military Guidelines on the Protection of, and Response to, Conflict-Related Sexual and Gender-Based Violence. Gender perspectives are also increasingly being incorporated in exercises. For example, NATO's 2015 crisis management exercise included, for the first time, a gender perspective as one of its objectives. These annual exercises are designed to practice the Alliance's crisis management procedures at the strategic-political level, involving civilian and military staff in Allied capitals, at NATO Headquarters and in both Strategic Commands.

Implementing the WPS agenda at NATO:

- The implementation of WPS Resolutions cuts across various divisions and governing bodies within NATO Headquarters as well as in the Strategic Commands. Together, these entities are responsible for monitoring and reporting the progress made by the Alliance. For this purpose, a Women, Peace and Security Task Force was established under the guidance and responsibility of the Special Representative for Women, Peace and Security.

- In sum, the mechanisms at NATO's disposal to implement the UNSC Resolutions are:

1. The Secretary General's Special Representative for Women, Peace and Security serves as the high-level focal point on all aspects of NATO's gender/WPS-related work. This position was created in 2012 and made permanent from September 2014. It is currently held by Clare Hutchinson;

2. A task force bringing together civilian and military staff across NATO Headquarters;

3. A gender adviser in the International Military Staff and an advisory committee of experts (NATO Committee on Gender Perspectives) on the military side, tasked with promoting gender mainstreaming in the design and implementation, monitoring and evaluation of policies, programs and military operations;

4. A working group led by Allied Command Operations to assess means to further incorporate the UNSCRs on WPS into operational planning and execution;

5. Gender advisers deployed at different levels of NATO's military command structure, including operational headquarters;

6. A number of relevant committees that develop and review specific and overall policy;

7. The NATO Science for Peace and Security (SPS) Program promotes concrete, practical cooperation on gender-related issues among NATO member and partner countries, through collaborative multi-year projects, training courses, study institutes and workshops;

8. The CSAP, to support and guide the work of WPS within NATO and advise on the integration of gender perspectives into NATO's core tasks.

*II.- Gender balance and diversity in NATO:*

NATO is an equal opportunities employer committed to valuing everyone as an individual. Gender balance and diversity efforts have been mainstreamed in NATO Headquarters (HQ) policies and practices since 2002. They aim at addressing issues such as imbalance in gender, age and national representation in the International Secretariat (IS) of NATO.

Recognising diversity means respecting and appreciating those who are different from ourselves. Today, there are approximately 1200 civilian IS members in NATO HQ. Another hundred civilians serve in the International Military Staff (IMS). They all operate under Civilian Personnel

Regulations, which provide that members of staff shall treat their colleagues and others, with whom they come into contact in the course of their duties, with respect and courtesy at all times. They shall not discriminate against them on the grounds of gender, race or ethnic origin, religion or belief, disability, age or sexual orientation.

Principles and priorities of gender and diversity at NATO HQ:

During the Prague Summit in November 2002, member countries tasked the IS to form a Task Force that would recommend to Council ways of improving gender balance and diversity in the NATO IS and civilian IMS workforce.

Under the direction of the Deputy Secretary General, the Task Force started work in February 2003. The first report proposed an Action Plan, which was noted by Foreign Ministers on 2nd June 2003. In consultation with national delegations, the IS and the IMS, the Task Force defined four guiding principles for actively pursuing a diversity policy at NATO HQ:

- Ensuring fairness in recruitment and promotion;

- Ensuring the high quality of NATO personnel;

- Respecting the diversity of all Alliance members; and

- Agreeing only to set goals and use methods that embody a reasonable challenge.

The Task Force therefore recommended a pragmatic approach with achievable goals. It focused on diversity issues that could be objectively defined and started its work by addressing the question of gender balance.

It agreed no quotas would be set since recruitment in NATO is merit-based, and proposed the following objectives:

- To increase the overall number of women employed in the IS;

- To increase the overall number of women applying (especially to A and C Grade positions);

- To increase the overall number of women in managerial positions.

Framework, monitoring and reporting:

a) A NATO-wide policy:

- To substantiate the above-mentioned decisions, NATO adopted a NATO-wide Equal Opportunities and Diversity Policy in 2003, applicable to the IS and civilian personnel in the IMS, as well as civilians in all NATO bodies and agencies.

- Separate policies against discrimination and harassment at work exist in NATO and several NATO bodies. Annual Progress Reports and Monitoring Reports are produced to outline achievements and trends and to put forward recommendations.

b) Some numbers:

- Currently 1178 people serve in the NATO IS of which 37.2% are women. Female personnel represent 31% of the A-grade staff and 22.5% of the senior management in NATO. Of the civilian personnel in the IMS, 43.9% are women. The PDF Library on this page provides a more detailed breakdown of gender, age and national representation in the NATO HQ's civilian workforce.

c) Mainstreaming diversity:

- A series of practical initiatives have been implemented in-house and continue to constitute a priority for NATO's services: the NATO Organizational Development and Recruitment services reviewed all job descriptions and vacancy announcements in order to ensure gender neutrality in their formulation. In addition, for senior posts at grade A.5 and above, an external assessment center may be used, which guarantees an additional level of culture-neutral professional assessment in line with NATO's merit-based recruitment principles.

- The Talent Management services work constantly on the personal and professional development of the NATO HQ workforce and provide specific training opportunities for women, as well as awareness-raising events for the entire IS. The team in the Personnel Support services is responsible for the general well-being of the NATO IS, whose health and balanced lifestyle are their priority.

- In 2004 the NATO Internship Program was established, allowing young graduates to bring to NATO HQ their share of diversity and enthusiasm. The success of the program led, in 2009, to its extension to all NATO bodies and agencies.

Action plans:

Bearing in mind the current demographic trends in NATO member states, and the vast number of international public and private institutions competing for quality candidates, it is crucial for the Organization to position itself well in order to remain, and for some to become, an employer of choice.

As the Organization changes in line with evolving political requirements and tasks, it is essential that NATO diversify qualifications and competencies of its workforce. The key to triggering sustained institutional change is mainstreaming the process of change, i.e., to fully weave it into the very fabric of the organisation. This is why, for instance, the first Action Plan covering the period 2007-2010 identified the three following objectives: to establish and maintain a NATO Diversity Framework and Policy; to improve the NATO work environment; and to promote and improve NATO's image as an employer of choice. For each one of these objectives, annual targets were set within the Action Plan and the Progress Reports monitor developments each year.

The next Action Plan should aim to shift work and efforts from diversity to inclusion. Diversity can be measured in numbers, but should not limit efforts to achieving balanced statistics. Rather, the aim would be to mainstream inclusion, which effectively means that efforts will be made to ensure that the diverse workforce will work well together.

*III.- Gender perspectives in NATO Armed Forces:*

Military operations in today's world require a diversity of qualifications and resources to ensure that peace and security are achieved and maintained. The complementary skills of both male and female personnel are essential for the effectiveness of NATO operations. The International Military Staff Office of the Gender Advisor and the NATO Committee on Gender Perspectives work to integrate a gender perspective into all aspects of NATO operations.

IMS office of the gender adviser:

The IMS Office of the Gender Advisor (IMS GENAD) reports directly to the Director General of the International Military Staff (DGIMS) and provides information and advice on gender issues, including the effective implementation of United Nations Security Council Resolution (UNSCR) 1325 and related Resolutions. It also serves as the Secretariat for the NATO Committee on Gender Perspectives (NCGP).

Among its responsibilities, IMS GENAD collects and disseminates information on the national policies relating to gender and the implementation of UNSCR 1325 and related Resolutions in NATO member and partner nations 'armed forces. Additionally, the Office facilitates dialogue with partner countries on relevant gender issues and liaises with international organisations and agencies concerned with the integration of a gender perspective into military operations.

NATO committee on Gender Perspectives:

The NATO Committee on Gender Perspectives (NCGP) promotes gender mainstreaming as a strategy for making the concerns and experiences of both women and men an integral dimension of the design, implementation, monitoring and evaluation of policies, programs and military operations.

By advising NATO's political and military leadership, as well as member nations, on gender-related issues and the implementation of UNSCR 1325 and related Resolutions, the NCGP contributes to operational effectiveness in line with Alliance objectives and priorities.

Other responsibilities of the NCGP include facilitating the exchange of information among NATO members on gender-related policies and gender mainstreaming, ensuring appropriate coordination on gender issues with the NATO Command Structure and NATO Headquarters, and collaborating with international organisations and agencies concerned with the integration of a gender perspective into military operations.

The NCGP is governed by an Executive Committee and supported by IMS GENAD. The Executive Committee is comprised of the Chair, the Chair-Elect, three Deputy Chairs and the IMS Gender Advisor, and must have at least one member of each gender. Both the Executive Committee and the Military Committee (NATO's senior military authority) can task the NCGP on specific gender-related issues.

Each NATO member and partner nation is entitled to designate one active duty officer of senior rank (or civilian equivalent) as a delegate to the NCGP. Delegates should be familiar with the latest national developments in gender approaches and tools for gender mainstreaming. They should also have knowledge of NATO and national policies relating to the implementation of UNSCR 1325 and related Resolutions.

Finally, non-NATO nations may be invited to contribute to the activities of the NCGP.


**Public diplomacy**

Communications and public diplomacy:

NATO communicates and develops programs to help raise awareness and understanding of the Alliance and Alliance-related issues and, ultimately, to foster support for, and trust in, the Organization. Since NATO is an intergovernmental organisation, individual member governments are also responsible for explaining their national defence and security policies as well as their role as members of the Alliance to their respective publics.

Role of communications and public diplomacy:

The overall aim of NATO's communications activities is to promote dialogue and understanding, while contributing to the public's knowledge of security issues and promoting public involvement in a continuous process of debate on security.

To do so, NATO engages with the media, develops communications and public diplomacy programs for selected groups including opinion leaders, academic and parliamentary groups, youth and educational circles. It seeks to reach audiences worldwide via its various platforms and social media activities. It also disseminates materials and implements programs and activities with external partners, while at the same time supporting the NATO Secretary General in his role as the principal spokesperson for the Alliance.

This drive to inform and engage with the public is reinforced by the knowledge that NATO is accountable to its member governments and their taxpayers who fund the Organization. As such, and in a spirit of transparency, it explains its policies, activities and functions.

a) Promoting security cooperation:

- Stimulating debate on NATO issues contributes to strengthening knowledge of the Alliance's goals and objectives. Many of NATO's information activities have an interactive, two-way nature, enabling the Organization to listen to and learn from the experience of its audiences, identify their concerns and fields of interest and answer to their questions. In Moscow and Kyiv for instance, NATO has set up information offices to increase the impact of its work and interact more frequently with its audiences.

- There are also information points in other partner countries and so-called "contact point embassies", which are NATO member country embassies located in partner countries that enable NATO to engage with local audiences.

b)  Types of activities:

- Today, the Alliance uses internet-based media and public engagement, in addition to traditional media, to build awareness of and support for NATO's evolving role, objectives and missions. In short, the Alliance employs a multi-faceted and integrated approach in communicating and engaging with the wider public.

- Over time, programs and policy have adapted to changes in the political and security environment, as well as to the technical innovations that have a direct impact on communication work. The communications services provided by NATO itself have also been reformed and restructured on numerous occasions to adapt to the different needs of the constantly evolving information environment, as well as to the needs of the security environment.

Working mechanisms:

The NAC and Secretary General are in charge of the overall direction of communications and public diplomacy programs for both the civilian and military sides of the Alliance.

The NATO Deputies Committee guides overall strategic communications on behalf of the NAC. Issue-specific NATO committees provide more detailed guidance, commenting on issues ranging from NATO maritime strategy to operations.

The Committee on Public Diplomacy (CPD) acts as an advisory body to the NAC on communication, media and public engagement issues. It makes recommendations to the NAC regarding how to encourage public understanding of, and support for, the goals of the Alliance.

At NATO Headquarters, members of the Public Diplomacy Division, who run communications and public diplomacy programs from within the International Staff, work closely with the International Military Staff and, more specifically, the Public Affairs and Strategic Communications Advisor to the Chairman of the Military Committee (MC). PDD also works with staff from the two strategic commands - Allied Command Operations (ACO) and Allied Command Transformation (ACT) - who communicate on operations, exercises and other activities under their purview. The interaction between the civilian and military side of the Alliance is key in ensuring a coherent and consistent approach to communications NATO-wide.

Evolution of communications:

The founding members of NATO understood the importance of informing public opinion. On 18 May 1950, the NAC issued a resolution in which it committed itself to: "Promote and coordinate public information in furtherance of the objectives of the Treaty while leaving responsibility for national programs to each country...". As early as August 1950, a modest NATO Information Service was set up and developed in the autumn with the nomination of a director. The service (similarly to the rest of the civilian organisation of the Alliance) did not receive a budget until July 1951. It effectively developed into an information service in 1952, with the establishment of an International

Staff headed by a Secretary General (March 1952), to which the information service was initially attached.

Later, in 1953, the Committee on Information and Cultural Relations (now the Committee on Public Diplomacy) was created. As such, from 1953, every mechanism was in place for the development of fully-fledged communications and information programs.

In 1956, the Report of the Three Wise Men stressed the overall importance of non-military cooperation and the need to develop unity within the Alliance. Cooperation in the information field was identified as one of the areas the Alliance should reinforce, stating: "The people of the member countries must know about NATO if they are to support it". To do so, it recommended: "The promotion of information about, and public understanding of NATO and the Atlantic Community should, in fact, be a joint endeavor by the Organization and its members".

## Science, research and technology

*I.- Science for Peace and Security Programme:*

The Science for Peace and Security (SPS) Program promotes dialogue and practical cooperation between NATO member states and partner nations based on scientific research, technological innovation and knowledge exchange. We will also highlight that the SPS Program offers funding, expert advice and support to tailor-made, security-relevant activities that respond to NATO's strategic objectives.

Introduction to the SPS program:

The SPS Program promotes security-related practical cooperation based on scientific research, innovation and knowledge exchange within NATO's wide network of partner countries.

It connects scientists, experts and officials from Allied and partner countries to address security challenges, such as cyber defence, counter-terrorism or defence against CBRN agents; to support NATO-led missions and operations; to foster the development of security-related advanced technologies such as sensors and detectors, nanotechnologies, unmanned aerial vehicles (UAVs); and to address human and social aspects of security such as the implementation of United Nations Security Council Resolution 1325 on Women, Peace and Security (UNSCR 1325).

In this regard, the SPS Program greatly benefits from the expertise of other NATO agencies, divisions and delegations, and bodies such as centers of excellence.

The Program provides the Alliance with a unique channel for non-military communication, including in situations or regions where other forms of dialogue are difficult to establish. It enables NATO to become actively involved in such regions, often serving as the first concrete link between NATO and a new partner.

The SPS Program has evolved continuously since its foundation in 1958. To this end, a comprehensive reorientation of the Program took place in 2013, which gave

SPS a renewed focus on larger-scale strategic activities beyond purely scientific cooperation.

SPS grant mechanism:

Funded by NATO's civil budget, the SPS Program supports collaboration through four established grant mechanisms: Multi-Year Research Projects, Advanced Research Workshops, Advanced Training Courses and Advanced Study Institutes. Interested applicants should develop proposals for activities that fit within one of these formats.

In the same sense, we will highlight that all activities funded within the framework of the SPS Program must follow the rules and regulations outlined in the SPS Program Management Handbooks.

To that end, interested parties submit an application for funding that must be led by project directors from at least one NATO Ally and one partner country. Any application must also directly address at least one of the SPS key priorities and have a clear link to security.

Once an application has been received by the SPS Program it will undergo a comprehensive evaluation and peer review process, taking into account expert, scientific and political guidance.

It must also be said that this process ensures that all SPS applications approved for funding have been thoroughly evaluated for their scientific merit and security impact by NATO experts, independent scientists and NATO nations themselves.

SPS support to NATO's political priorities:

SPS flagship projects contribute to several of NATO's key partnership initiatives and priorities and have been reflected as deliverables in various NATO Summit documents.

A) Defence and Related Security Capacity Building (DCB) Initiative:

- The DCB Initiative was launched at the 2014 NATO Summit in Wales in order to reinforce NATO's commitment to partners by providing support to nations requesting defence capacity assistance from NATO. The SPS Program is currently supporting the DCB packages for Iraq, Jordan and the Republic of Moldova.

    1. In Iraq, security forces were trained in the area of Counter-Improvised Explosive Devices (C-IED) and were provided with related specialist equipment. Iraq's C-IED operations support humanitarian efforts to return displaced populations safely to their homes. Furthermore, an advanced level, hands-on cyber defence training course was organised for Iraqi system / network administrators to directly respond to requirements of the Iraqi authorities.

    2. In Jordan, the SPS Program supported the development of a national cyber defence strategy. It thereby significantly enhanced Jordan's cyber defence posture and established a Computer Emergency Response Team (CERT) for the Jordanian Armed Forces (JAF). SPS

further supported the JAF in the domain of C-IED through tailor-made training courses that have been designed and implemented in collaboration with the NATO C-IED Center of Excellence in Spain.

3. In the Republic of Moldova, an SPS multi-year project established a cyber defence laboratory to serve as a training center for civil servants of the defence and security relevant institutions. System and network administrators of the Moldovan Ministry of Defence also received a comprehensive cyber defence training. Furthermore, a multi-year project launched in 2016 is supporting the Moldovan government and civil society actors in creating a multi-agency national strategy to implement UN Security Council Resolution 1325 on Women, Peace and Security.

B) Projecting stability in NATO's neighborhood through practical cooperation:

- At the 2016 Warsaw Summit, NATO leaders emphasised their commitment to contributing more to the efforts of the international community in projecting stability and strengthening security beyond NATO borders. Through dialogue and practical cooperation with partner nations, the SPS Program actively contributes to these efforts.

- It thereby assumes a balanced and flexible 360-degree approach to help address the security challenges to the east and south of the Alliance, including terrorism.

1) Enhanced explosive remnants of war (ERW) detection and access capability in Egypt:

    - This project, launched in 2014, provides Egypt with an enhanced operational detection and clearance capability for ERW. Provision of this capability will enhance the safety of Egyptian deminers, reducing the number of casualties from ERW clearance and improving their individual confidence and credibility. This will have an immediate effect on the safety and security of the local population, lowering the threat from ERW and releasing land for economic development.

2) Next Generation Incident Command System in the Western Balkans:

    - This flagship project, supported by the SPS Program and the US Department of Homeland Security, Science & Technology Department, is developing and implementing a system to facilitate the coordination among first responders and improve civil emergency management across the Western Balkans. The new technology will allow responders to share all kinds of information about an incident, including the GPS location or images, via mobile devices. This will maximise real-time situational awareness and help find a coordinated, appropriate response to natural or man-made disasters.

3) CBRN first responders live agent training:

- The overarching goal of this live agent train-the-trainer course hosted by the Joint CBRN Defence Center of Excellence (JCBRN CoE) in Vyškov, Czech Republic, was to enable 17 first responders from Egypt, Jordan and Tunisia to survey, monitor and manage the consequences of a CBRN incident. Experts from the Organization for the Prohibition of Chemical Weapons (OPCW) reinforced the JCBRN CoE and provided instructor support. The training was designed to assist the partner nations to improve their civil emergency plans, complement national training systems and improve cooperation between first responders.

C) Comprehensive Assistance Package (CAP) for Ukraine:

- At their meeting in Warsaw on 9 July 2016, the Heads of State and Government of the NATO-Ukraine Commission endorsed the CAP for Ukraine. The objective of the Package is to consolidate and enhance NATO's assistance for Ukraine in order to make the country's defence and security institutions more effective, efficient and accountable. As part of the CAP, the SPS Program implemented several activities in Ukraine's priority areas of cooperation.

➢ A multinational telemedicine system enables medical specialists to engage in major disasters or incidents across national borders. Portable medical kits allow first responders at the scene to connect to the system to receive expert advice from medical specialists in case of an emergency, even in remote areas. Through the use of modern communications technologies, an international network of medical specialists is able to assess patients, diagnose them and provide real-time recommendations. This allows the right aid and care to reach those who need it most quickly, with the potential to save many lives.

➢ Another SPS project to support humanitarian demining in Ukraine enhanced the capacity of the State Emergency Service of Ukraine (SESU) in undertaking demining operations in the eastern part of the country. The overall aim was to safeguard the civilian population within areas affected by the conflict and allow the return of displaced persons. The project will be complemented by a multi-year initiative to develop an innovative 3D landmine detection radar.

*II.- NATO Science and Technology Organization:*

The NATO Science and Technology Organization (STO) delivers innovation, advice and scientific solutions to meet the Alliance's ever-changing needs. It ensures NATO maintains its military and technological edge to face current and future security challenges.

Main tasks and responsibilities:

The STO generates and exploits a leading-edge science and technology program of work, delivering timely results and advice that advance the defence capabilities of Allies, partners and NATO in support of the core tasks of collective defence, crisis management and cooperative security.

It also supports decisions made at both national and NATO level by providing advice to the North Atlantic Council and national leadership.

The STO achieves its mission by nurturing a community of more than 5,000 actively engaged scientists. The STO network draws upon the expertise of more than 200,000 people in Allied and partner nations.

Structure:

The STO is governed by the NATO Science and Technology Board (STB). The Board administers the STO's scientific and technical committees and its three executive bodies: the Center for Maritime Research and Experimentation (CMRE) in La Spezia, Italy; the Collaboration Support Office in Paris, France; and the Office of the Chief Scientist at NATO Headquarters in Brussels, Belgium.

The Chief Scientist is the chairman of the STB and the senior science advisor to the North Atlantic Council.

The scientific and technical committees, composed of members from national and NATO bodies, direct and execute NATO's collaborative science and technology activities.

The CMRE organisations and conducts scientific research and technology development, centered on the maritime domain, delivering innovative solutions to address the Alliance's defence and security needs.

The CMRE conducts hands-on scientific and engineering research for the direct benefit of both NATO and such customers as research entities and industry. The Center operates NATO's two research vessels that enable science and technology solutions to be explored and developed at sea. This allows unique and specialised research to be conducted in core areas of interest for NATO. The CMRE's engineering capability enables rapid exploitation of concept prototypes for use in trials and military experiments. The Center also has a scientific and engineering knowledge base composed of a dedicated science platform and publications, for use across NATO.

Evolution:

To safeguard Alliance freedom and shared values, it is of critical importance for NATO and its partner nations to maintain the edge in defence and security. Discovering, developing and utilising advanced knowledge and cutting-edge science and technology is fundamental to maintaining the technological edge that has enabled Alliance forces to succeed across the full spectrum of operations over the past decades.

The STO was created through the amalgamation of the Research and Technology Organization and the NATO

Undersea Research Center. These bodies were brought together following a decision at the 2010 NATO Summit in Lisbon to reform the NATO agency structure.

The STO, along with its predecessor organisations, has been instrumental in enabling that success, both within the nations and for NATO itself.

In the same sense, we will highlight that by providing a critical venue for knowledge development and delivery, the STO remains committed to its foundational principle: bringing together subject matter experts from across the scientific spectrum with military end users in order to inform decision-makers on emerging challenges and opportunities, and to ensure the technological advantage of the Alliance and its partners.

*III.- NATO's role in energy security:*

The disruption of energy supply could affect the security of societies of Allies and partners, and have an impact on NATO's military operations.

While these issues are primarily the responsibility of national governments, NATO Allies continue to consult on energy security and further develop NATO's capacity to contribute to energy security, concentrating on areas where it can add value. NATO seeks to enhance its strategic awareness of energy developments with security implications; develop its competence in supporting the protection of critical energy infrastructure; and work towards significantly improving the energy supply for the military.

Activities:

a) Enhancing strategic awareness of the security implications of energy developments

- While NATO is not an energy institution, energy developments affect the international security environment and can have far-reaching security implications for some Allies. A stable and reliable energy supply, the diversification of routes, suppliers and energy resources, and the interconnectivity of energy networks are of critical importance and increase resilience.

- NATO closely follows relevant energy trends and developments, and seeks to raise its strategic awareness in this area. This includes consultations on energy security among Allies and partner countries, enhancing intelligence-sharing and assessments, and expanding links with relevant international organisations, such as the International Energy Agency and the European Union.

- NATO also organises specific events, such as workshops, table-top exercises and briefings by external experts. Of particular importance in this regard are the North Atlantic Council's bi-annual seminars on regional and global energy developments, as well as the annual Energy Security Strategic Awareness Course at the NATO School in Oberammergau.

b) Supporting the protection of critical energy infrastructure

- All countries are reliant on energy infrastructure, including in the maritime domain, on which their energy security and prosperity depend. Energy infrastructure is also one of the most vulnerable assets, especially in areas of conflict. Since infrastructure networks extend beyond borders, attacks on complex energy infrastructure by hostile states, terrorists or hacktivists can have repercussions across regions. For this reason, NATO seeks to increase its competence in supporting the protection of critical energy infrastructure, mainly through training and exercises.

- Protecting energy infrastructure is primarily a national responsibility. However, since NATO forces are dependent on civilian energy infrastructure, it is important that Allies strengthen their infrastructure to take account of NATO's requirements. Moreover, NATO organisations exercises and exchanges best practices with partner countries, many of which are important energy producers or transit countries, and with other international institutions and the private sector.

- By protecting important sea lanes, NATO's counter-piracy operations have also made an indirect contribution to energy security. Moreover, NATO is also supporting national authorities in enhancing their resilience against energy supply disruptions that could affect national and collective defence.

c) Enhancing energy supply for the military

- The high fuel demand of combat forces can diminish their performance, increase their vulnerability, and may require the diverting of combat forces to protect supply lines. Hence, increased energy efficiency could offer benefits in terms of combat power and agility. A significant step forward in this area was the adoption of NATO's "Green Defence" framework in February 2014. It seeks to make NATO more operationally effective through changes in the use of energy, while saving resources and enhancing environmental sustainability.

- NATO's own work in this regard focuses on reducing the consumption of fossil fuel in deployed force infrastructure (i.e. camps), resulting in more autonomy, a lesser logistical burden and a smaller environmental footprint. Allies are also reviewing the fuel supply chain for NATO forces in a more demanding security environment.

Evolution:

At the Bucharest Summit in 2008, Allies noted a report on "NATO's Role in Energy Security", which identified guiding principles and outlined options and recommendations for further activities. These were reiterated at subsequent summits, while at the same time giving NATO's role clearer focus and direction.

The 2010 Strategic Concept, the setting up of an Energy Security Section in the Emerging Security Challenges Division at NATO Headquarters that same year, and the accreditation of the NATO Energy Security Center of Excellence in Lithuania in 2012 were major milestones in this process.

The decision of Allies to "integrate… energy security considerations in NATO's policies and activities" (2010 Lisbon Summit Declaration) also meant the need for NATO to reflect energy security in its education and training efforts, as well as in its exercise scenarios. Since then, several exercises have included energy-related developments, and several training courses have been stood up, both nationally and at the NATO School in Oberammergau, Germany.

At the Brussels Summit in July 2018, Allies underlined the important role energy security plays in their common security and that it is essential to ensure that the members of the Alliance are not vulnerable to political or economic manipulation. In November 2019, Allies agreed to a set of recommendations on consolidating NATO's role in energy security, which included in particular a stronger focus on how to ensure a viable fuel supply to the military.

In the years to come, NATO Allies will continue to seek diversification in energy supplies, further enhance the strategic dialogue both among Allies and with partner countries, offer more education and training opportunities, and deepen ties with other international organisations, academia and the private sector .

Work on enhancing the resilience of energy infrastructure, notably in hybrid scenarios, will also be given greater attention. With increased awareness of energy risks, enhanced competence to support infrastructure protection, and a more energy-efficient military, NATO will be better prepared to respond to the security challenges of the 21st century.

*IV.- Environment - NATO's stake:*

NATO recognises that it faces many environmental challenges. In particular, the Alliance is working to reduce the environmental effects of military activities and to respond to security challenges emanating from the environment.

The Alliance first recognised the natural environmental challenges facing the international community in 1969, when it established the Committee on the Challenges of Modern Society (CCMS). Until its merger with the NATO Science for Peace and Security (SPS) Program in 2006, the CCMS provided a unique forum for NATO and its partner countries to share knowledge and experience on social, health and environmental matters, both in the civilian and military sectors .

Over the years, Allied countries have established several NATO groups to address environmental challenges from various angles.

NATO's current activities related to the natural environment include:

- o protecting the environment from damaging effects of military operations;

- o promoting environmentally friendly management practices in training areas and during operations;

- o adapting military assets to a hostile physical environment;

- o preparing for and responding to natural and man-made disasters;

- o addressing the impact of climate change;

- o educating NATO's officers on all aspects of environmental challenges;

- o supporting partner countries in building local capabilities;

- o enhancing energy efficiency and fossil fuel independence; and

- o building environmentally friendly infrastructures.

All these activities fall under two broad categories:

1) Environmental protection: Protecting the physical and natural environment from the harmful and detrimental impact of military activities.

2) Environmental security: Addressing security challenges emanating from the physical and natural environment.

1.- Environmental protection:

Military activities often have an adverse effect on the environments in which they occur. Damage to the environment from these activities can threaten livelihoods and habitats, and thus breed instability. Part of NATO's responsibility is to protect the physical and natural environments where operations and training take place.

Since the 1960s environmental experts have argued that the military should adopt measures to protect the physical and natural environment from harmful and detrimental effects of its activities.

Environmental degradation can cause social and economic instability and new tensions, whereas the preservation of the environment during a military operation can enhance stabilisation and foster lasting security. Hence, minimising environmental damage during training and military operations is of great importance for the overall success of the mission.

NATO member countries are aware of the environmental challenges during military operations and they have adopted rules and regulations to protect the environment. NATO's measures range from safeguarding hazardous materials (including fuels and oils), treating waste water, reducing fossil fuel consumption and managing waste to putting environmental management systems in place during NATO-led activities. In line with these objectives, NATO has been facilitating the integration of environmental protection measures into all NATO-led military activities.

a) Policy and standards (including evolution and mechanisms paragraph):

- NATO started to develop its environmental protection policy in the late 1970s when NATO expert groups and processes were established to address environmental challenges, resulting in a number of guidelines and standards. At this time, NATO's policy states that NATO-led forces "must strive to respect environmental principles and policies under all conditions".

- Currently, two dedicated NATO groups are addressing environmental protection while promoting cooperation and standardisation among NATO and partner countries, as well as among different NATO bodies and international organisations that regularly attend as observers:

    a. the Environmental Protection Working Group (EPWG) (under the Military Committee Joint Standardization Board that reports to the Military Committee)

    b. the Specialist Team on Energy Efficiency and Environmental Protection (STEEEP) (under the Maritime Capability Group "Ship Design and Maritime Mobility" that reports through the NATO Naval

Armaments Group to the Conference of National Armament Directors).

- The EPWG aims to reduce possible harmful impacts of military activities on the environment by developing NATO policies, standardisation documents, guidelines and best practices in the planning and implementation of operations and exercises.

- The ST/EEEP aims to integrate environmental protection and energy efficiency regulations into technical requirements and specifications for armaments, equipment and materials on ships, and for the ship to shore interface in the Allied and partner nations' naval forces.

- Two decades of activities by expert groups have paved the way for the overarching policy document MC 469 on "NATO Military Principles and Policies for Environmental Protection", of which the first version was agreed by the NATO Military Committee in 2003, and an updated version was agreed upon in October 2011.

- This document describes the responsibilities of military commanders for environmental protection during the preparation and execution of military activities. Further, it recognises the need for "a harmonisation of environmental principles and policies for all NATO-led military activities". It also instructs NATO commanders to apply "best practicable and feasible environmental protection measures", thus aiming at reducing the environmental impact caused by military activity. The MC 469 has been complemented with several other NATO EP Standardization Documents (STANAG) and Allied Joint Environmental Protection Publications (AJEPP), all focused on protection the environment during NATO-led military activities. These include the following:

- STANAG 7141 Joint NATO Doctrine for Environmental Protection During NATO-led Military Activities (AJEPP-4)

- STANAG 2510 Joint NATO Waste Management Requirements During NATO-led Military Activities (AJEPP-5)

- STANAG 2582 Environmental Protection Best Practices and Standards for Military Camps in NAT-led Military Activities (AJEPP-2)

- STANAG 2583 Environmental Management System in NATO Operations (AJEPP-3)

- STANAG 6500 NATO Camp Environmental File During NATO-led Operations

     - STANAG 2594 Best Environmental
Protection Practices for Sustainability
of Military Training Areas (AJEPP-7)

b) Training:

- In order to ensure compliance with such standards, forces must receive appropriate environmental protection training. While such training is primarily a national responsibility, it is NATO's ambition to provide common environmental protection and energy efficiency education to Allies' forces. It is necessary to embed environmental protection awareness into the daily routine of military personnel and increase their personal responsibility in this field. To advance this objective, NATO has designated staff officers for the implementation of environmental protection at strategic, operational and tactical levels. As well, NATO School Oberammergau and the Military Engineering Center of Excellence (MILENG COE) provide environmental protection courses and instruction as part of their curriculum.

c) Research and Development:

- NATO's Science and Technology Organization (STO) promotes and conducts scientific research on military-specific technical challenges, some of which are related to environmental issues. To this end, STO technical/scientific sub-committees, composed of experts from NATO and nations, look for "greener solutions" by conducting studies and research resulting in scientific reports. STO's activities include noise reduction and "greener ammunition". The STO's Center for Maritime Research and Experimentation (CMRE) located in La Spezia, Italy, conducts research to quantify the impact of the environment on operations, and vice versa. One extensive CMRE study resulted in a better understanding on how marine mammals can be affected by sonar systems.

- Based on the results, NATO developed the "Code of Conduct for the Use of Active Sonar to Ensure the Protection of Marine Mammals within the Framework of Alliance Maritime Activities" (MC-0547). STO's Collaborative Network is supported by the Collaboration Support Office, located in Paris, France.

- Within the context of NATO's Science for Peace and Security (SPS) Program, environmental protection experts across NATO and partner nations have been active in the development of policy and technical solutions to the reduction of the environmental and energy footprint on NATO-led activities.

- One such advanced research workshop consisted of the development of a NATO Camp Closure Handbook and a Sustainable Camp Model.

- The model enables operational planners to better understand the impact of operations on water, waste and energy consumption and provides technical solutions aimed at a reduction in the environmental and energy footprint of operations.

d)  Collaborative Approach:

- NATO's Environmental community has been active in their cooperative efforts with other international organisations, to include the UN and EU. This collaborative approach also includes discussions with industry, academia and governmental agencies.

2.- Environmental security:

Based on a broad definition of security that recognises the importance of political, economic, social and environmental factors, NATO is addressing security challenges emanating from the environment. This includes extreme weather conditions, depletion of natural resources, pollution and so on - factors that can ultimately lead to disasters, regional tensions and violence.

The Alliance is looking closely at how to best address environmental risks to security in general as well as those that directly impact military activities. For example, environmental factors can affect energy supplies to both populations and military operations, making energy security a major topic of concern. Helping partner countries clean up aging and dangerous stockpiles of weapons, ammunition and unexploded remnants of war that pose a risk to people and the environment is yet another area of work.

NATO is currently conducting these initiatives via its Science for Peace and Security (SPS) program, the Euro-Atlantic Disaster Response Coordination Center (EADRCC) and Partnership for Peace Trust Fund projects. It is considering enhancing its efforts in this area, with a focus on civil emergencies, energy efficiency and renewable power, and on consulting with relevant international organisations and experts on NATO's stake in climate change.

a) Building international cooperation:

- Since 1969, NATO's SPS Program has supported cooperative activities that tackle environmental security issues, including those that are related to defence, in NATO countries. Since the SPS Program opened up to partner countries in the 1990s, partners listed environmental security as a top priority, requesting NATO's support for cooperative activities to address those issues that threaten the security of their country and beyond.

- In order to better coordinate its activities, NATO joined in 2004 five other international agencies under the Environment and Security (ENVSEC) Initiative[1] to address environmental issues that threaten security in four vulnerable regions. The regions are South east Europe, Eastern Europe, South Caucasus and Central Asia.

- As a first step, ENVSEC facilitated regional meetings with relevant stakeholders (especially, experts, non-governmental organisations authorities, governmental authorities and international donors) to consult and agree on regional maps highlighting priority issues that are a threat to security. As a second step ENVSEC raised fund to address the identified issues, The SPS program mainly support capability building through projects that helped partner countries with equipment, consumables, travel, training and stipends.

b) Boosting emergency response:

- The Alliance is also actively engaged in coordinating civil emergency planning and response to environmental disasters. It does this principally through the Euro-Atlantic Disaster Response Coordination Center (EARDCC) that was launched following the earthquake disaster in Turkey and Greece at the end of the 1990s.

- Talking at the UN Climate Change Conference in Copenhagen, NATO's former Secretary General Anders Fogh Rasmussen highlighted that, with the growing impact of climate change, the demand upon the military as "first responder to natural disasters" was likely to grow. I have urged Allies to consider how to optimise the Alliance's contribution in that area. With the aim to increase the understanding, NATO organised consultations and scenario building exercises involving military and civilian experts, partly supported by the SPS Program. Consequently, under NATO's current Secretary General Jens Stoltenberg the dialogue with other international organisations has been enhanced with a focus on how NATO and its armed forces could better adapt to the challenge of an increasing number of natural disasters.

c) Energy security and Critical Energy Infrastructure Protection:

- With increasingly unpredictable natural disasters, such as earthquakes, severe floods and storms that cause disruptions to infrastructure, environmental factors have a growing potential to affect energy security, a challenge NATO is becoming aware of. Most NATO members and partners rely on energy supplies from abroad, sent through pipelines and cables that cross many borders. Allies and partners, therefore, need to work together to develop ways of reducing the threat of disruptions, including those caused by environmental events.

- At the Strasbourg/Kehl Summit in April 2009, Allies said they will "consult on the most immediate risks in the field of energy security". They said they would continue to implement the recommendations proposed at the 2008 Bucharest Summit, namely to share information, advance international and regional cooperation, develop consequence management, and help protect critical infrastructure.

- Projects that focus on the link between energy infrastructure and environmental security have been supported by the SPS Program since early 2000. An example is the multi-year project "Chernobyl Dust Model" that is helping Ukraine to develop a realistic 3D model of the radioactive dust that is leaking from the damaged sacrophage at the Chernobyl Nuclear Power site.

- This will not only increase the safety of the workers of the New Safety Confinement, but also helps international experts understand the challenges of measurements and monitoring of contaminated areas.

d) Energy efficiency in the military (Smart Energy):

- Recognising the increasing need of fuel in operations, causing security issues for fuel convoys and armed forces, NATO started in 2011 a Smart Energy initiative bringing together NATO stakeholders and national experts from the public and private sector. Heads of State and Government declared in Wales in 2014 that NATO will "[...] continue to work towards significantly improving the energy efficiency of our military forces, and in this regard we note the Green Defence Framework".

e) Helping partners reduces environmental hazards through disarmament:

- Through NATO's Partnership for Peace Trust Fund projects, the Alliance helps partner countries reduce their aging weapon stockpiles, clean up deteriorating rocket fuel, clear land contaminated by unexploded remnants of war and safely store ammunition. While the central aim is to help post-Soviet countries disarm and reform their militaries, these projects also reduce the risks posed by these dangerous materials to the environment and the people in surrounding areas.

f) Raising awareness and information-sharing:

- Communicating the security implications of environmental issues to political leaders and decision-makers is another area where the Alliance plays a major role. For instance, it makes sure that members and partners alike have the knowledge and skills needed to mitigate climate change and adapt to its effects.

# APPENDIX 1

## THE NORTH ATLANTIC TREATY
### Washington D.C., 4 April 1949

The Parties to this Treaty reaffirm their faith in the purposes and principles of the Charter of the United Nations and their desire to live in peace with all peoples and all governments.

They are determined to safeguard the freedom, common heritage and civilization of their peoples, founded on the principles of democracy, individual liberty and the rule of law.

They seek to promote stability and well-being in the North Atlantic area. They are resolved to unite their efforts for collective defense and for the preservation of peace and security.

They therefore agree to this North Atlantic Treaty:

## ARTICLE 1

The Parties undertake, as set forth in the Charter of the United Nations, to settle any international disputes in which they may be involved by peaceful means in such a manner that international peace and security, and justice, are not endangered, and to refrain in their international relations from the threat or use of force in any manner inconsistent with the purposes of the United Nations.

# ARTICLE 2

The Parties will contribute toward the further development of peaceful and friendly international relations by strengthening their free institutions, by bringing about a better understanding of the principles upon which these institutions are founded, and by promoting conditions of stability and well-being. They will seek to eliminate conflict in their international economic policies and will encourage economic collaboration between any or all of them.

# ARTICLE 3

In order more effectively to achieve the objectives of this Treaty, the Parties, separately and jointly, by means of continuous and effective self-help and mutual aid, will maintain and develop their individual and collective capacity to resist armed attack.

# ARTICLE 4

The Parties will consult together whenever, in the opinion of any of them, the territorial integrity, political independence or security of any of the Parties is threatened.

# ARTICLE 5

The Parties agree that an armed attack against one or more of them in Europe or North America shall be considered an attack against them all; and consequently they agree that, if such an armed attack occurs, each of them, in exercise of the right of individual or collective self-defense recognized by Article 51 of the Charter of the United Nations, will assist the Party or Parties so attacked by taking forthwith, individually and in concert with the other Parties, such action as it deems necessary, including the use of armed force, to restore and maintain the security of the North Atlantic area. Any such armed attack and all measures taken as a result thereof shall immediately be reported to the Security Council. Such measures shall be terminated when the Security Council has taken the measures necessary to restore and maintain international peace and security.

## ARTICLE 6[1]

For the purpose of Article 5 an armed attack on one or more of the Parties is deemed to include an armed attack on the territory of any of the Parties in Europe or North America, on the Algerian departments of France[2], on the occupation forces of any Party in Europe, on the islands under the jurisdiction of any Party in the North Atlantic area north of the Tropic of Cancer or on the vessels or aircraft in this area of any of the Parties.

## ARTICLE 7

This Treaty does not affect, and shall not be interpreted as affecting, in any way the rights and obligations under the Charter of the Parties which are members of the United Nations, or the primary responsibility of the Security Council for the maintenance of international peace and security.

## ARTICLE 8

Each Party declares that none of the international engagements now in force between it and any other of the Parties or any third state is in conflict with the provisions of this Treaty, and undertakes not to enter into any international engagement in conflict with this Treaty.

## ARTICLE 9

The Parties hereby establish a council, on which each of them shall be represented, to consider matters concerning the implementation of this Treaty. The council shall be so organized as to be able to meet promptly at any time. The council shall set up such subsidiary bodies as may be necessary; in particular it shall establish immediately a defense committee which shall recommend measures for the implementation of Articles 3 and 5.

## ARTICLE 10

The Parties may, by unanimous agreement, invite any other European state in a position to further the principles of this Treaty and to contribute to the security of the North Atlantic area to accede to this Treaty. Any state so invited may become a party to the Treaty by depositing its instrument of accession with the Government of the United States of America. The Government of the United States of America will inform each of the Parties of the deposit of each such instrument of accession.

## ARTICLE 11

This Treaty shall be ratified and its provisions carried out by the Parties in accordance with their respective constitutional processes. The instruments of ratification shall be deposited as soon as possible with the Government of the United States of America, which will notify all the other signatories of each deposit. The Treaty shall enter into force between the states which have ratified it as soon as the ratifications of the majority of the signatories, including the ratifications of Belgium, Canada, France, Luxembourg, the Netherlands, the United Kingdom and the United States, have been deposited and shall come into effect with respect to other states on the date of the deposit of their ratifications [3].

## ARTICLE 12

After the Treaty has been in force for ten years, or at any time thereafter, the Parties shall, if any of them so requests, consult together for the purpose of reviewing the Treaty, having regard for the factors then affecting peace and security in the North Atlantic area, including the development of universal as well as regional arrangements under the Charter of the United Nations for the maintenance of international peace and security.

## ARTICLE 13

After the Treaty has been in force for twenty years, any Party may cease to be a Party one year after its notice of denunciation has been given to the Government of the United States of America, which will inform the Governments of the other Parties of the deposit of each notice of denunciation.

## ARTICLE 14

This Treaty, of which the English and French texts are equally authentic, shall be deposited in the archives of the Government of the United States of America. Duly certified copies thereof will be transmitted by that Government to the Governments of the other signatories.

1)  The definition of the territories to which Article 5
    applies was revised by Article 2 of the Protocol to
    the North Atlantic Treaty on the accession of Greece
    and Turkey signed on 22 October 1951.

2)  On January 16, 1963, the North Atlantic Council
    noted that insofar as the former Algerian
    Departments of France were concerned, the relevant
    clauses of this Treaty had become inapplicable as
    from July 3, 1962.

3)  The Treaty came into force on 24 August 1949, after
    the deposition of the ratifications of all signatory
    states.

www.ingramcontent.com/pod-product-compliance
Lightning Source LLC
Chambersburg PA
CBHW061500120726
48001CB00004B/1162